AVOIDING POLICY FAILURE: A WORKABLE APPROACH

AVOIDING POLICY FAILURE:
A WORKABLE APPROACH

Steven E. Wallis

EMERGENT™

PUBLICATIONS

Avoiding Policy Failure: A Workable Approach
Written by: Steven E. Wallis

Library of Congress Control Number:
      2011940984

ISBN: 978-0-9842165-0-5

Copyright © 2011
Emergent Publications,
3810 N. 188th Ave, Litchfield Park, AZ 85340, USA

Printed in the United States of America

CONTENTS

APPLYING NEW TOOLS OF COMPLEXITY SCIENCE TO SOLVE OLD PROBLEMS OF POLICY

HOW TO STRUCTURE SUCCESSFUL INTERNATIONAL GOVERNMENTAL ORGANIZATIONS: CREATE RELIABLE TREATIES, AND IMPROVE INTERNATIONAL RELATIONS

CHOOSING A MILITARY POLICY THE HARD WAY

WHEN GOOD EFFORTS CAUSE POOR RESULTS: HOW TO IDENTIFY PROBLEMATIC ECONOMIC POLICIES BEFORE THEY FAIL

THE FUTURE OF POLICY

ABOUT THE AUTHOR

Steven E. Wallis, Ph.D., is the director of the Foundation for the Advancement of Social Theory (FAST). A lifelong learner and scholar-practitioner, Steve earned his doctorate from Fielding Graduate University in 2006. He has a decade of experience as an organizational development consultant in Northern California and a broad range of interdisciplinary interests. Currently, he is a Fellow at the Institute for Social Innovation at Fielding and serves on the editorial board of the *Integral Review*. He also teaches at Capella University and mentors doctoral candidates. Recently, he served as Editor for a special issue of the *Integral Review* focused on *Emerging Perspectives of Metatheory and Theory*. He is the editor for *Cybernetics and Systems theory in Management*, a book that offers a wide range of innovative perspectives.

ABOUT FAST

The Foundation for the Advancement of Social Theory (FAST) is a long-view, transdisciplinary project dedicated to accelerating the advancement of social theory to fulfill the promise of the social sciences by providing theory that is unarguably effective in practical application. FAST is striving to identify new and more useful paths for validating social theory, benchmarking the advancement of theory, and supporting the falsification and application of effective theory in social systems. From a metapolicy perspective, FAST is here to quantify the internal coherence/integrity of policy and certify the potential effectiveness of that policy. Where others analyze the issues and create policy, we analyze policy using rigorous methodologies to quantify the internal logics of that policy. We also provide specific, workable, recommendations for improving policy. This unique service supports governments, IGOs, NGOs, agencies, corporations, and policy analysts in their efforts to create policy that will be more effective in practical application. Another way we work to accomplish our goal is to support our Research Fellows in their efforts to conduct rigorous critical and integrative metatheoretical research. FAST provides education and mentoring to help new and established scholars to conduct effective research and increase their rate of publication in the academic literature. Please, visit us at: http://ProjectFAST.org.

ACKNOWLEDGEMENTS

Some parts of this book were derived in part or whole from: "Towards the development of more robust policy models" as published in the *Integral Review*, 6 (1). http://integral-review.org/

Among many others, I would like to thank Greg Daneke, Gianfranco Minati, Ilya Garber, Katrina Rogers, and Robert Silverman for their constructive suggestions that resulted in a more clear and coherent publication. Of course, any errors remain strictly my own.

PREFACE

This book is targeted at graduate students, scholars, analysts, and decision makers who are teaching, learning, and working in the policy field. Specifically, those who are seeking better ways of understanding, creating, choosing, and implementing more effective policy. As such, this book may be applied as a textbook for learning innovative and useful methods, as a scholarly study showing how complexity thinking may be applied to policies, and as a starting point for creating new insights into metapolicy analysis.

Policy is important because each policy serves as a guide for decisions of organizational, corporate, community, national, and global importance. The text of each policy may be seen as a "lynchpin" in the process of research and practice that determines the effectiveness, efficiency, and the validity of decisions that may cost trillions of dollars (for the national budget) thousands of lives (in military conflicts) and touch the lives of billions of human beings (for economic and environmental issues). With better policy, our society will make better decisions that could save thousands of lives, billions of dollars, and reduce needless suffering.

Generally, studies have focused on the goals, reasons, actions, and results of policy. In an important and innovative contrast, the present investigation is focused on the policy text, itself. Here, the text is seen as a logical cognitive construct that is amenable to objective analysis and quantification. The text represents the policy maker's understanding of the situation, not the goals desired or the actions required. That difference will allow us to make policy decisions that will lead to more policy successes than the many policy failures of the past.

As I write this book, we approach the tenth anniversary of September eleventh, 2001. Only now are we beginning to extricate ourselves from two long and costly wars. It is easy to imagine that we might have completed those wars much earlier if we had possessed more effective military policy. It is even possible that we might have avoided those wars completely by avoiding the unexpected terrorist attacks that triggered them. To do so, we would have needed policies of unprecedented quality in their ability to help us understand our world and predict world events.

Recent advances in critical metapolicy suggest new approaches for analysis based on insights from complexity theory. Specifically, that we can quantify the complexity and the co-causal relationship between the propositions within a policy. And, critically, that there is a correlation between the quantifiable structure of a policy and the effectiveness of that policy in practical application. It has been suggested in the literature that we can use methods such as propositional analysis (PA) to determine the effectiveness of a policy prior to implementation based on the policy text. Such an approach would enable scholars to develop more effective policy and provides a new tool for practitioners to choose between competing policies.

In this book, I test that assertion by applying PA to six policies in three comparative case studies. Cases include military policy, economic policy, and international policy. Because of the great difficulty associated with finding policies that were effective (let alone comparable cases), these studies may be seen as somewhat obscure. I certainly invite all readers to join in an effort to find additional cases for more comparisons.

In each case comparison, the quantified structure of the policy is compared with the historical consequences of implementing the policy. Generally, the results of the study support the assertion. I found that policies with higher levels of structure (higher internal integrity and greater complexity) tend to be more effective in practical application. And, conversely, policies of lower complexity and less internal integrity tend to be less effective. Additional insights are also discussed along with implications for future research and application. Some important next steps for this line of research would be to conduct additional case comparative studies as well as larger scale, statistical analyses. The usefulness of this methodology across a range of policy fields suggests that it is generalizable across most, perhaps all, areas of policy interest.

APPLYING NEW TOOLS OF COMPLEXITY SCIENCE TO SOLVE OLD PROBLEMS OF POLICY

Introduction and Overview

This book aims to change the way we evaluate policy prior to its implementation. This will impact the policy process by providing practitioners, politicians, and the general public with a new way to compare policy proposals that is workable and effective. This approach will also support policy analysts in their efforts to develop more effective policies and help scholars to advance policy science more readily.

In this book, I will begin by touching on the relationship between theory, policy, and the many problems faced in our socioeconomic lives. In brief, saying that many of our issues are due to a lack of effective theory and policy. Key to that meta-problem is that the social sciences have not had a reliable method for choosing between policies except to implement them. That process is, of course, difficult and expensive. It also means that the process of choosing a policy is more influenced by political spin, rather than effective analysis.

Following that overview, I will introduce propositional analysis (PA), an innovative method derived from complexity theory. In this, PA is conceptually similar to the innovative DYSAM approach which uses metamodeling to highlight emerging properties within systems (Minati, 2010). Using PA, it is possible to determine the potential usefulness of a theory based on the diversity of concepts and the co-causal logical structures found within the theory. One use of PA is to quantify the extent to which a logical structure may be understood as a non-linear system. In this, PA may be understood as a tool to identify relations between concepts, rather than to categorize those concepts.

In a previous paper (Wallis, 2010d), I suggested that PA could be applied to the analysis of policy. The central hypothesis of this book essentially asks if there is indeed a correlation between that quantifiable structural integrity of a policy and the usefulness of that policy in practical application. To put it another way, this book asks if we can look at the texts of two policy documents and determine which one is more likely to work in practical application?

In the following three chapters, I present three case comparison studies designed to answer that question. In each case, I present two policies and analyze their complexity and structural integrity. I then compare each policy with the historical effects from implementing the policy. In Chapter Two, I compare the Charter of the United Nations with the Covenant of the League of Nations. One policy led rapidly to a failed organization and another global conflict; the other policy led to a more sustainable organization and relative peace and prosperity. In Chapter Three, I compare two military policies from about 1870. One military policy led to victory; the other led to surprising defeat. In Chapter Four, I compare two economic policies from the 1980s. Here, one policy led to economic stability and higher levels of employment while the other led to wild fluctuations of inflation and unemployment.

What I find is that the initial hypothesis is supported—it is possible to determine the potential effectiveness of a policy before it is applied. This, of course, has powerful implications for policy research, development, analysis, and conversations in the public sphere.

In the final chapter, I summarize the finding and discuss some implications for policy practitioners, policy researchers, scholars, and provide insights for metapolicy thinkers and metatheoreticians.

One important learning for this book is that rigorous complexity thinking can be applied effectively to interpret and evaluate policy documents—just as complexity thinking can be used to interpret and advise actions of managers, analysts and world leaders. More specifically, PA can be used to objectively analyze two policies and determine which one will be more effective in practical application. This sets the stage for rapid advancements in the way policy is developed and radical improvements in the way policy is chosen and applied. Such improvements, in turn, open the door to unprecedented peace, prosperity, and socioeconomic stability.

The Failure of Social Theory

Increasingly, scholars and practitioners are questioning the usefulness of theory in the natural sciences (Smolin, 2006) and the social sciences. Indeed, the persistence of the many problems of the world has made it increasingly clear that the social sciences do not provide effective tools for understanding or guiding our individual and organizational activity. This limitation is reflected in the spotty success of economics (Dubin, 1978, citing Rapoport), the failure of social change theory (Appelbaum, 1970; Boudon, 1986), high failure rates in the application of Total Quality Management (MacIntosh & MacLean, 1999), frequent failure of organization development culture change (Smith, 2003), failure of organizational theory (Burrell, 1997), and theories of bureaucracy (Bernier & Hafsi, 2007). Even the currently popular field of management theory is still striving for legitimacy (Arbaugh, 2008: 5) and may cause more problems than it solves—such as the collapse of Enron (Ghoshal, 2005). Our ability to make theory is in doubt (Kessler, 2001) as we wonder why our theories are not advancing (Starbuck, 2003). To the point where some have sought to abandon theory (e.g., Shotter, 2005).

This lack of ability to create effective theory is reflected in our inability to develop effective policy and practice and sends us back to the drawing board in an effort to concoct new theories from combinations of old (e.g., Ugiabe & Obetoh, 2011). Our nations are beset by a multitude of problems including war, poverty, injustice, and economic collapse. Problems that would be minimized if we were in possession of adequate theories and policies. Our inability to develop effective public policy models has left the promise of the social sciences "largely unfulfilled" (Spicer, 1998) at a time when "The complex and difficult global environment has overwhelmed, exasperated and saddened many observers" (Dennard, Richardson & Morçöl, 2008: 17). For example, our current economic problems may be blamed on "policy shortcomings" (Wroughton & Kaiser, 2008).

For every dimension of our society, we need policy that is unarguably effective, not "an answer that is clear, simple, and wrong" (Menken, 2009: 1). If we had policies that worked effectively, we would know what steps to take to alleviate these problems. We would not spend so much time and effort arguing about which policy (if any policy) to implement. For example, the "war on drugs" started by Nixon. Under Regan the funding was greatly increased. His policy, however, was simplistic. Essentially, he claimed that drugs were dangerous so more enforcement was needed. Today, despite great expense, massive incarceration, and an overloaded court system, the problem seems to be as bad as it was before—and in many ways worse (Baum, 1996). Indeed, it may be that attempting to solve complex problems with simple policies leads to large unanticipated negative outcomes (Wallis, 2010d).

A more complex, well-integrated policy might be expected to be several times more effective. For example, rather than an approach that is focused only on law enforcement, we might instead have an approach which includes community health, scientific research, balancing the benefits of drugs with their costs to individuals and communities, and works towards recovery (without excluding the potential for prosecution (Wallis, 2010d). Such a complex and integrated policy might have us in the same place we are today—but with less cost. Or, for the same cost, we might expect to have minimized problems associated with drug use.

Here, I am not proposing a fuzzy utopia; rather, I am saying that by using more effective conceptual tools we will be better able to solve serious social problems.

The limits of classic policy theory (including those developed by Baumgartner, Jones, Kingdon, Sabatier, and Jenkins-Smith) seems to have "come to the end of its line of development" (John, 2003: 482). The need for better policy propels agencies into a "metapolicy environment" (Boschken, 1994: 308) in hopes of finding ways of making policy that are better than Kingdon's "policy soup" (White, 1994: 869) which might be conceptually similar to the "garbage can" model of decision making. Even an empirical approach to program evaluation "has not led to successful policies or programs" (Schmidt, Scanlon & Bell, 1979: 1). Indeed, "The history of policy development is marked by unanticipated consequences resulting from insufficient appreciation of large-scale system dynamics that characterize policy systems" (Albritton, 1994: 159).

Alternative approaches to policy such as "muddling through" (Lindblom, 2010) do not appear to be useful; particularly given the failure of that approach in Viet Nam, Iraq, and the current financial crisis (Scott Jr., 2010). Without a rigorous form of metapolicy analysis, the voting public falls prey to questionable claims that a proposed policy "makes sense" (Bastedo, 2005). Worse, conflicting claims of sensemaking seems to mobilize and antagonize partisans causing useless arguments in the public sphere.

Nor are scholars immune from policy confusion. On an academic level we risk becoming, "mesmerized by the technical brilliance, peerless morphology, and operational complexity of the latest model" (Der Derian, 1996: 87). For example, we may consider the supposed benefits of computer models. However, they "are tools that can be used and easily abused" (Richardson, 2008: 51). In short, while it is good try new methods, we must take them with a grain of salt. We cannot continue to apply useless methods with great rigor; we must develop meta-methods for evaluating our methods.

In his critique of the modern policy process, Sabatier (1999) calls for better theories to improve the process of policy creation. And, to some extent, there is an influx of new ideas and theories from sociology and organizational studies (Bastedo, 2007). New approaches to policy have been suggested from the study of complexity theory. For example, "triangulation" is suggested as a way to coordinate multiple methods of analysis (Roe, 1998). "Full spectrum analysis" suggests the need to carefully analyze a wide range of information (Mathieson, 2004). Other, more general, approaches have also been suggested (Elliott & Kiel, 1999). However, they have not yet been proven effective. So, a new method is needed.

DeLeon (1999) suggests that a systems approach might be a useful way to better understand policy. Indeed, it appears that "the time is ripe for a systematic evaluation of our metatheoretical assumptions" (Lamborn, 1997: 212). Or, more plainly, we need to develop a better understanding of how we develop policy if we are to develop better policy. The present book answers those calls by introducing a new method of critical metatheory from complexity studies and organizational studies. Using such methods to compare and improve policies is expected to engender significant advances in the field of policy analysis.

What's In a Policy?

Within the policy process, the many streams of problem emergence, political wrangling, and policy analysis result in the creation of a policy document—an explicit written report that presents the shared understanding of what the problem is and how it might be addressed within a limited window of opportunity (Kingdon, 1997).

The policy delineates the processes, level of description, and context of events that are considered meaningful to the organizational system. Within this policy document, causal propositions provide an implicit or explicit logic model. A well-written policy document will have an explicit logic model. Older, and less coherently constructed, policy documents will have implicit logic models.

The logic model "represents stakeholders' best guess or theory for what will be most effective" (Hernandez & Hodges, 2001: 10). Unfortunately, "The criteria for comparing frameworks are not well developed" (Schlager, 1999: 252). In the present book, I suggest that those criteria are poorly understood because the literature has focused on the relationship between the policy and the "real world," rather than focusing at the logical relationships between the propositions within the model, itself. In this book, I will move toward repairing and restoring the balance between the empirical and the logical.

This topic is essentially the study of metapolicy (how to better understand how to make better policy) and, "is a major subject in urgent need of work as a main issue of redesigning governance for the 21st century" (Dror, 1994: 21). Rather than idiographic (as is often found in historical or policy studies), this kind of analysis may be understood as nomothetic (law-like) where, "The focus of this design is on relationships of variables… [not] characteristics" (Albritton, 1994: 161). The above insights from philosophy, complexity theory, and structural logic suggest a new perspective from the nascent field of critical metatheory (Wallis, 2010b).

Previously, the term metapolicy has been used to refer to a "policy about policies" (Kerr, 1976: 351), an explanation of the social context within which policy is developed (Jacobs, 1995), the collaboration of various groups (Walters, Aydelotte & Miller, 2000), or policy networks (Detomasi, 2007: 321). The present use is closer to a discussion of the "structure" of policy problems (Hoppe, 2002). In the present book, I will use the term metapolicy analysis specifically to refer to the analysis of policy text. The investigation and analysis of policies and their internal logics might be broadly considered a form of design science, policy science, or evaluation research (Van de Ven, 2007: 278).

A policy model may be understood as a mental model, schema, theory, or (more colloquially), a lens or point of view. The policy model acts as a computer program, metaphor, filter, or sense-making device. For example, if a nation has a policy model that represents immigrants as invaders, that nation might take action to fight back against those immigrants. In contrast, holding a different lens of policy, a nation that sees immigrants as useful contributors to society would welcome immigrants with open arms. For each approach, there will be some outcomes that are somewhat predictable, and some results that are less so.

While "policy studies rely heavily on analysis of systems" (Albritton, 1994: 166) those analyses have left tacit the system inherent to the policy's internal logics. To find the logics within the policy text, the present book will use a form of hermeneutics which involves reading a text and deconstructing that text into its constituent parts for deeper analysis (Bentz & Shapiro, 1998; Lyotard, 1984). This approach is similar to studies of institutional logics which include the, "assumptions, values, beliefs, and rules" individuals use to "provide meaning to their social reality" (Thornton & Ocasio, 2008: 101). For the present book, I will avoid the form of hermeneutics which is focused on personal reflection (e.g., Wight, 2011). Instead, I will focus on forms of logical, co-casual, relationships between the constituent parts, which I will rigorously quantify.

With rigor, we have science. (Wallis, 2010b) Without rigor, we have speculation, which leads to argument and conflict about how to reduce global problems. The present book adds a new and important tool to the policy scholar's kit by providing a way to be rigorous about logics. I should note here that this book is focused on the policy text. Concerns about the origins of those texts (empirical studies, speculation, experience, etc.), while also of importance, are considered secondary to this specific form of analysis. It is equally important to note here that the present approach may be used to support those forms of analysis; thus allowing researchers to quantify and compare policies, origins, implementations, and results for a greater understanding of policy science.

Textual hermeneutics has an advantage over reflective hermeneutics because the textual approach may be applied more completely. For a negative example, if two scholars reflect on a political situation, each will likely perceive different parts of the problem (because the entire situation is much too large for any one person to fully comprehend). Complete analysis becomes possible when we are studying text because the text is simpler than the whole situation and yet serves as a reasonable abstraction of the situation. And, in the event of a disagreement, both scholars may refer back to the original text.

Emerging Insights—Complexity and Structure of Policies

For this book, a policy is not "what we want;" that is more properly called a goal. Neither is policy "what we do;" for that would merely be describing a tautological relationship between actions and plans (Etzioni, 2010). Similarly, this book avoids the research, analysis, and insight needed for the creation of a policy as well as the requirements for successful implementation of that policy; those topics are well studied elsewhere (e.g., Dennard *et al.*, 2008). Here, instead, a policy is better understood as a *cognitive structure representing how we understand the world so that we know what do to achieve our goals.* The question becomes one of how might we rigorously measure the text of a policy?

One field, closely related to policy, is the study of International Relations (IR). Within that field, a debate is now raging around metatheory and its potential to help or harm the field (Freire, 2011). Freire, along with the field of IR, rightly eschews the unhelpful view of metatheory as a fuzzy or overarching theory created by speculative or specious reasoning and instead accepts the view that a metatheory may be used to evaluate theory. He identifies, among others, concerns within the field that metatheory might increase the complexity of IR and should be avoided for that reason. Freire, however, argues cogently against such critiques noting, "the best critiques of metatheory are inherently metatheoretical" (Freire, 2011: 25). Thus, metatheory is an inescapable perspective that may serve to improve theory—if we can develop more effective forms of metatheoretical analysis.

As one form of analysis, McLaughlin and Jordan (1999) suggest investigating the logic model as a set of hypotheses, logical statements, or propositions. However, they do not go much beyond this recommendation. So, the present book, in developing more effective analyses, may be understood as an extension of their insightful work. While systems approaches have been applied for analyzing situations and creating IR policy (Chittick, 2006; de Green, 1993; Hayden, 2005) that systemic thinking has not been applied to study the policies, themselves.

In his cogent discussion on scientific realism, Chernoff (2002) highlights some of the confusion encountered in the process of creating theories of IR. While acknowledging the benefits of testing theories by experimentation and/or practical application, Chernoff also builds a strong case for these kinds of theory noting,

"Theories should be eliminated from further consideration if they fail to meet a priori requirements like logical consistency" (p. 195).

Where previously we have not had a rigorous method to quantify and thus compare the logical consistency of policies, the development of PA (more on this below) provides a tool to measure the coherence of a cognitive structure such as a theory or policy. This seems a particularly promising area of investigation based on my studies into the texts of theories where I found a correlation between the structure of a theory and the usefulness of that theory in practical application (Wallis, 2010a). That work also empirically demonstrates how the complexity of the theory increases as the theory evolves from a low level of usefulness to a higher level of usefulness.

Structure is also reflected in the internal logics of a policy. Importantly, this is not a process of asking if the policy *seems* "logical" or "makes sense." Because, if we ask multiple scholars to read the same policy, not all will agree that the policy is logical or sensible. A more useful approach is to deconstruct the structures of logic to identify what specific forms of logic are used within a policy. That way, it is possible to quantify the qualitative forms of logic and so arrive at a more accurate and objective understanding of how logical a policy might be.

There are at least five forms of logical structure that are relevant to this approach. The first four are: atomistic (a form of truth claim), linear (simple casual), circular (tautological) and branching (where one cause results in multiple effects). These first four forms of logical structure, however, are of limited use. Atomistic truth claims are inherently weak (Metcalfe, 2004), and circular arguments are generally accepted as unacceptable. Complexity thinking agues against the linear logics as does Stinchcombe (1987). Branching seems to compound confusion rather than alleviate it.

The fifth structure of logic is seen in the proposition: "More A *and* more B cause more C." These kinds of propositions are *concatenated*. That is to say, there are multiple causes that are evidenced in the resultant effect (Kaplan, 1964; Van de Ven, 2007).

This concatenated structure is conceptually similar to Bateson's (1979) idea of "double description" where multiple streams of information are combined to suggest a new (third) form of information that is more useful than the previous two. An example of double description is binocular vision (where the extra sense of depth is added). Bateson shows how these double descriptions create an extra dimension of understanding that is of a different logical type. This approach may also be considered conceptually similar to a dialectical process of thesis, antithesis, and synthesis. The concatenated aspect may be understood as an emergent aspect, or synthesis of the other two.

In brief, concatenated aspects (e.g., "C" in the above paragraph) are privileged in this form of analysis because they are better understood (and so make *more* sense) than aspects connected by other logical structures (e.g., atomistic, circular, linear, branching). Thus, a concatenated structure may also be understood as representing a space where new understandings emerge. In short, measuring a conceptual system's number of concatenated aspects allows us to measure a conceptual system's potential to engender emergence.

Methods—Propositional Analysis

For the present book, each policy will be analyzed using propositional analysis (PA). Ideally, this method should rely on the exact wording of the original text. For the present analysis, I have modified some of those words slightly to clarify the casual relations and to reduce the need for long explanations that would not fit into the present book. Those changes are not meant to alter the meaning of the original text in any way.

The PA analyses of those texts provide objective indicators of the structure, or internal coherence, of the policy. One indicator is the Complexity, the variety of concepts or aspects, within the policy. Here, Complexity (with a capital C) will be used to refer to the calculated diversity of ideas within a policy document. Another indicator is Robustness. My studies show that theories which are more Robust (like those of physics and mathematics) will prove more effective in practice, while less robust theories (such as those commonly found in the social sciences) will prove less effective in practice (Wallis, 2010a).

This use of the term "robust" differs from other meanings. It does not mean, "insensitive to uncertainty about the future" (Lempert & Schlesinger, 2000: 391). Nor is it used in the more colloquial senses of "strong" or "unchanging." Instead, Robustness is a specific and objective measure of the relatedness between propositions within the theory. In short, PA identifies what percentage of a policy's aspects are well understood (compared with the total number of aspects in the policy). In that approach, this book uses the term Robustness in a way that is more consistent with its usage in physics and mathematics in referring to theories (or laws) that are amenable to algebraic manipulation.

PA has been used to analyze cognitive structures in a variety of fields including complexity theory (Wallis, 2011a), policy models (Wallis, 2010d), physics (Wallis, 2010a), management and organizations (Wallis, 2009a, 2011b), ethics (Wallis, 2010c), entrepreneurship (Wallis, 2009b), and cross disciplinary theories of everything (Wallis, 2008b). This broad reach, from practical to abstract and from physics to social sciences, suggests the unique and useful capacity of this method.

In a sense, PA asks how we might understand the policy as a system. It answers that question by analyzing the level of integration between the aspects of the policy, or the extent to which the logical propositions within the policy are related to one another. For policy analysis, PA is applied in six steps:

1. Identify a specific policy text
2. Identify all causal propositions within the text (preferably a specific policy model or similar concise, yet authoritative, representation of the policy)
3. Link propositions according to related aspects (creating a diagram is very helpful here)
4. Quantify the total number of aspects to find the Complexity
5. Identify and quantify the concatenated aspects

6. Divide the number of concatenated aspects by the total number of aspects to find the Robustness (a ratio between zero and one).

For an abstract example, let us say we have a policy consisting of the following propositions: A is true; B is true; more A causes more C; more B causes more D; more D and more C cause more E. In such a model, there are five aspects (A, B, C, D, and E). Therefore, the Complexity of the policy is C=5.

Of those five, only E is concatenated (increases in D *and* increases in C cause increases in E). This allows us to find the ratio of well-understood aspects to poorly understood aspects of R=0.20 (the result of one concatenated aspect divided by five total aspects).

Very generally, theories of the social sciences are found to be around or below this low level of Robustness. In what may be more than a simple coincidence, organizational change models (based on such theories) seem to be effective about 20% of the time (e.g., Dekkers, 2008; MacIntosh *et al.*, 1999; Smith, 2003). Therefore, it may be suggested that the Robustness of a model provides some indicator of a model's effectiveness and so this policy would be effective about 20% of the time (or, perhaps, may be expected to achieve only 20% of its goals). More study is needed to confirm or falsify this general observation.

Conclusion and Overview of Analysis

In the present chapter, I have touched on the problems of the world as well as our inability to solve those problems with the existing tools of social theory and policy. Looking at the underlying philosophy of theory and policy, the root of that problem appears to be an over reliance on empirical data and a corresponding lack of understanding about the forms of logic used to describe the relationships between the data within the policy text. New approaches have been suggested using complexity theory. Specifically, propositional analysis has emerged as a rigorous tool for analyzing policy models so that they can be legitimately, objectively, and rigorously deconstructed and tested for the level of Complexity and Robustness that is related to the usefulness of the policy.

Casey & Brugha (2005) characterize policy as a "wicked complex problem" They note, "From the shaman's reading entrails and the Oracle at Delphi... people seek to determine what the future will be. Modern attempts at long-range economic and weather forecasts have fared little better than their predecessors, due to the nonlinear nature of the phenomena" (Casey *et al.*, 2005: 44-45). The present book suggests that the best way to address a nonlinear problem in the world is with a policy built of non-linear (co-casual) logics. Propositional analysis is a rigorous tool that measures the nonlinearity of a policy; and, through that measurement, allows us to choose between two policies to decide which one is more likely to be usefully applied in practical application.

In the following three chapters, I will use PA to analyze six policies—then compare the Complexity and Robustness of those policies with the historical results of their implementation. This will provide us with some insights into the potential efficacy of PA as an analytical tool and point the way for more effective policy creation and evaluation.

Chapter Two includes a comparison of two constitutional documents. One is from the United Nations and the other from the League of Nations. Chapter Three compares two military policies. One is from the French and the other from the Prussians. The fourth chapter compares two economic policies. One from Australia and the other from the Netherlands.

Because of the need to compare policies with their results, I use a comparable-cases strategy (Lijphart, 1975) for these studies. That approach calls for two cases with similar independent variables and dissimilar dependent variables. That approach has been applied widely including studies of organizations on an international level (e.g., Yan & Gray, 1994) and comparisons of economic stabilization policies (e.g., Courchene, 1999). In a broad sense, this may be understood as a form of content analysis (Hjørland, 2002; Hood & Wilson, 2002; Semler, 2001). It may also (in a limited sense), be understood as a kind of cliometric analysis (Faust, 2005; Meehl, 1992, 2002, 2004).

In each chapter, propositional analysis will be used to determine the Complexity and Robustness of two policies. The Complexity and Robustness of each policy will then be compared with measures from outside the policy such as the longevity and success of the policy. Finding correlations between the structure of the policy and the results of the policy will validate the PA method and how we may create better policies and more successful organizations.

HOW TO STRUCTURE SUCCESSFUL INTERNATIONAL GOVERNMENTAL ORGANIZATIONS: CREATE RELIABLE TREATIES, AND IMPROVE INTERNATIONAL RELATIONS

Introduction

This chapter represents a form of regime analysis through a form of constitutional analysis. This may also be understood as relating to institutional theory as well as International Relations theory and the creation of effective international treaties because the founding constitution of an international governmental organization (IGO) may also be considered a kind of treaty (Schrijver, 2006).

A constitution is a blueprint for the organization: a formalized set of norms. In that document, are instructions for how the organization is to function—its internal workings and interrelationships between units, reporting hierarchies, budgeting functions, etc. There are also indicators of how that organization is to understand and engage the world around it. The constitution (as a formal, purposefully negotiated, expression of the organization's culture) also serves as a schema or theory—a lens through which the organization's members view and understand the world (Legro, 1997: 36).

In the history of our planet, two international organizations have been created that are simultaneously unique and comparable. The League of Nations (League) and the United Nations (UN) were both formed to address a vast range of global issues. The failure of the League led to the founding of the longer-lived UN. Despite 65 years of operation, the UN is not always considered successful in addressing important global issues. So, there is room for improving the policy and practice of this organization.

These two institutions have long been the subject of study and comparison. For example, (Diehl, 2005: 3) claims that the League "clearly" failed because of a "failure of will" on the part of the great nations and because of the, "unwieldy requirements for concerted action" in the League Covenant. Despite his claim of clarity, other scholars present different reasons for the failure of the League. For example, "the redistribution of power in favor of the smaller nations and the free-rider problem caused by the non-binding nature of the League's decisions" (Sobel, 1994: 173) and, "the failure to provide adequate security guarantees for its members" (Eloranta, 2010: 27).

For the present analysis, it is not necessary to rebut the claims made by these researchers. It is enough to note that they make different claims for the causes of the League's failure. Indeed, of the six reasons set out by these three scholars, there appear to be more differences than similarities. Because there is no general agreement as to why the League failed, it seems reasonable to conclude that we really do not know why. This lack of understanding may explain why the UN has not been able to significantly improve its ability to achieve its most vaunted goals despite the dedicated efforts of many talented individuals.

Analysis

This study uses PA to determine the Complexity and Robustness of two constitutional documents. Those results will then be compared with historical data to develop new insights into the creation of more effective and useful constitutions. The discussion will develop suggestions (and cautions) for improving the UN and for the creation of more effective IGOs.

Independent Variable—Two Constitutions

Given the size of the founding documents, a complete detailed analysis is beyond the scope of the present chapter. Therefore, I will focus on the preamble for each document as a condensed version or abstraction. In short, the preamble serves as a type of policy model or logic model that serves to represent the key ideas of the complete document.

Dependent Variable—Longevity

While one may argue about the relative effectiveness of each organization, and/or the historical context under which the organization operated, it is difficult to argue with the idea that the organization which exists is more effective than an organization which does not. Therefore, longevity seems to be a reasonable approximation of organizational success.

The Charter for the League of Nations was signed in 1919 following World War One. The first meeting was held in 1920. The last formal meeting of the League was in 1946 when it was officially dissolved. Charitably, we can say that the League lasted for 26 years between signing and dissolution (although its effective life was shorter—less than 20 years). Its fall is even said to have begun after only eleven years in 1931 when the League failed to intervene against Japan's invasion of Manchuria (Sobel, 1994: 177).

The United Nations charter was signed in 1945 following World War Two. The first meeting was held in 1946 and they are still meeting today. The United Nations has lasted for 65 years, and it appears to be viable for the immediately foreseeable future. The clear difference between these two organizations on this dimension suggests that the lifespan of the organization is a good dependent variable for this case-comparison study.

Complexity and Robustness

The Covenant of the League of Nations

The constitution for the League is referred to as the Covenant and is over 4000 words. A copy of the complete Covenant may be found at: http://avalon.law.yale.edu/20th_century/leagcov.asp. The introduction, Article One, from http://avalon.law.yale.edu/20th_century/leagcov.asp#art1. The introduction to the League begins with five concepts. 1. Nations should not resort to war, 2. Nations should have open, just, and honorable relations, 3. Nations should maintain justice and respect treaties, 4. Nations should accept that international law determines how nations should act, 5. The previous four ideas will lead to international cooperation, peace, and security. These five concepts and their causal relationships are presented in Figure 2-1.

Using PA, we can see from this diagram that there are five aspects. Therefore, we can say that the Covenant has a

Figure 2-1 *Causal Diagram of the Covenant of the League of Nations, Article 1*

Complexity of C=5. There appear to be four causal aspects to the Covenant and one resulting, (concatenated) aspect. Therefore, the Robustness of the Covenant appears to be R=0.20 (the result of one divided by five).

The Robustness of R=0.20 seems to be generally similar with the low Robustness of many theories of the social sciences (e.g., Wallis, 2009b). Worse, there are some relationships between the aspects of the Covenant that appear inherently weak. For example, the idea that less war will lead to more peace may be understood as a useless tautology. Also, the idea that international law should be accepted as determining conduct is something of a tautological recursion. First, because the entire document is an international treaty and so (presumably) has the force of law. Second, because if a nation's conduct does not respect or uphold the law, that behavior throws into question the extent to which the law really serves as a law to determine behavior. Therefore, the *functional*

Complexity and Robustness of this diagram may be lower than suggested by its 0.20 score.

It is also possible to "unpack" the aspects of Figure 2-1. For example, Aspect #2 might be described as three aspects by separating out "justice," "honor," and "relationships." Therefore, the functional Complexity of Article One might be higher than suggested by Figure 2-1. However, the authors presented the Covenant in this way, and it is important to stay true to their original intentions (Ritzer, 1990, 2001). Otherwise, we are not so much analyzing their Covenant as re-interpreting it. Such a revisionist approach would render the present analysis less than optimal by "contaminating" the independent variable.

The Charter for the United Nations

The constitution for the UN is referred to as the Charter and is nearly 9000 words long (Schrijver, 2006). A copy of the Charter and is available at: http://www.un.org/en/documents/charter/index.shtml. The Preamble is available at http://www.un.org/en/documents/charter/preamble.shtml. The aspects from the Preamble and their causal relationships are presented in Figure 2-2.

In Figure 2-2, there are 13 total aspects (one in each box). Thus, the Complexity of the Charter is C=13. Only one of these aspects is concatenated (number 13 is seen to be caused by 11 and 12). Using propositional analysis, Robustness of the charter is 0.08 (the result of one divided by 13).

While the Charter has a very low level of Robustness (because there is only one concatenated aspect) the structure of the model is not completely hopeless. There

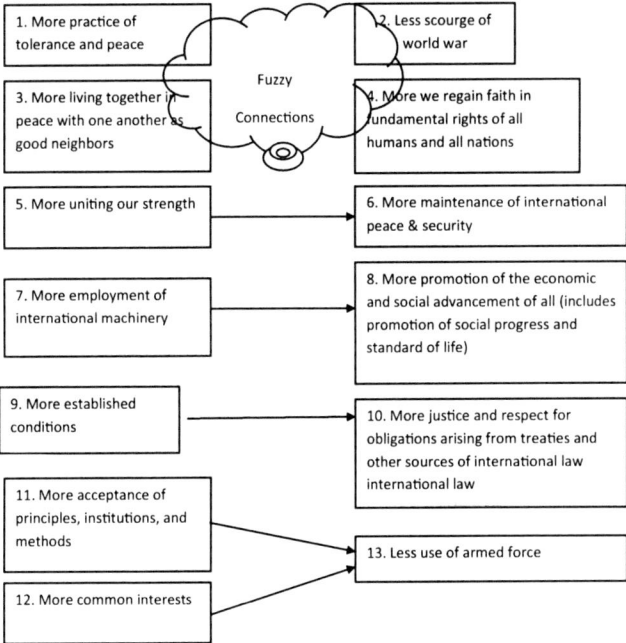

Figure 2-2 *Causal Diagram of Aspects of the Preamble to the United Nations Charter*

are, three linear causal connections that are not accounted for in propositional analysis. Also, there are some poorly defined causal relationships (between aspects 1, 2, 3, & 4 in Figure 2-2). The extent to which these aspects may be more effectively structured may suggest a higher functional Robustness than initially indicated using propositional analysis, alone. Although, it should be noted, the critical integrative metatheoretical perspective holds that linear causal connections are not of significant value.

Comparison and Conversation

The creation of the two organizations has been well documented (e.g., Annan, 2000; Goodrich, 1947; Walters, 1986). World leaders gathered to create both the League and the UN. We may reasonably infer that those leaders were experts in their fields enjoying the benefits of well-honed political intuition and supported by world-class research. Both constitutions were created by a coalition of victors following global conflict. Also, both may be understood as a kind of international treaty (Schrijver, 2006)—a political process of negotiation between national entities. The similarities of origins suggest that this test of creation does not have significant bearing on the differences in longevity of these two IGOs.

Complexity and Robustness

For the present analysis, the size of the founding documents offers a simple indication of their relative Complexity. The League's Covenant contains slightly over 4,000 words while the UN Charter contains a little less than 9,000 words. So, it may be seen that the UN Charter is more than twice as large as the League's Covenant, thus indicating greater complexity (small "c"—thus vague and poorly understood) and hinting at greater Complexity (large "C"—carefully calculated).

The League's Covenant contains only five aspects, while the UN Charter contains 13. This shows that the Charter contains more than twice the number of aspects as the Covenant. Note that the number of aspects correlates closely to the number of words in the complete document. This suggests that we are on the right track as far as using the introductory portion of each document to reflect the Complexity of the document as a whole. A higher level of Complexity suggests a conceptual construct is more mature and hence may be more useful in practical application (Commons, Trudeau, Stein, Richards & Krause, 1998).

The present comparison suggests that the Complexity of the constitution (which serves as a theory, policy, schema, set of norms) is related to the ability of the organization to perceive important changes in its environment (François, 2008) and to take effective action, adapt to its environment, and even to survive (Rhodes, 2008: 361). It stands to reason that the more options held by an organization, the wider its potential range of strategic decisions, and better understanding of environmental factors.

Additionally, it should be noted that the more complex UN evolved out of the simpler League. This fits with the evidence showing that more complex theories may evolve from simpler theories—and become more useful as they do (Wallis, 2010a). Although, that same study suggests that increases in Complexity is an intermediate step that may lead to increasing Robustness and still greater levels of usefulness.

When I compare the Robustness of the two documents, the differences are not so clear. The League Covenant has a Robustness of R=0.20 while the UN Charter is R=0.08. From one perspective, we might say that the Robustness of the League is more than twice the Robustness of the UN. In analyses of theories of the social sciences, however, this kind of difference does not appear to be meaningful because they are both at the low end of the scale.

Metaphorically, the sailboat of our policy is driven by the wind of Robustness and the tide of Complexity. While the wind is strong, we may go where we will—almost heedless of the tide. When the wind is very weak, we are moved about by the wind and tide equally.

The low level of Robustness for both organizational documents provides new insight as to why each organization has failed to achieve their highest goals of global peace, eradication of poverty, universal human rights, and other vaunted objectives. Although the norms of both organizations are strongly dedicated to the eradication of war, those norms are weak because they are not will integrated with the other norms. Such poorly integrated norms crumble more easily than carefully integrated ones (Wallis, 2010c).

Resources

While money, as a resource, is not the absolute indicator of organizational success it certainly plays a role. In the present case study, it appears that the League grossly underestimated the funding requirements for effective operation (a mistake that may also be related to the creation of a relatively ineffective constitution). For its first ten years of operation, the League budget (excluding peacekeeping forces) varied between $3.4 million and $5.2 million per year. In today's dollars, accounting for inflation, this works out to $38 to $69 million. Or, on average, about $56 million per year (Budget, 1930).

In contrast, for the first ten years of operation, the annual budget of the United Nations ranged from $25 million to $85 million (Singer, 1959). Adjusting for inflation that works out to $288 to $703 million. So, in short, one reason for the success of the UN over the League may be traced to the level of funding. The UN continues to survive because it receives approximately ten times the financial resources as the League—in addition to the complexity of its constitution.

This observation suggests that a more complex constitution requires more resources to support the larger staff and more complex operations. Or, from another perspective, large amounts of money spent to support a simple plan may well be wasted. This is because the simple plan will not account for the myriad complexities of the situation and so it becomes less likely that the resources will be allocated where they will do the most good. For example, the "war on drugs" received much better funding under Regan than under Nixon or Carter—yet does not seem to have been any more successful.

Conclusions

What the present analysis did not find was an expected correlation between the Robustness of the constitution itself, and the longevity of the organization. This is not seen as a great problem because both the League Covenant and the UN Charter both have very low levels of Robustness (R=0.20 and R=0.08, respectively). Given those low levels, we may reasonably infer that it is possible to create IGOs with greatly improved constitutions that will be much more effective in their work by founding them with highly Robust constitutions.

While we may reasonably accept that the survival of the UN is related to the Complexity of its constitution and the wealth of available resources, a question arises as to the relative benefits of those factors. To what extent does that funding represent a level of support that is necessary and sufficient to support the Complexity of the organization? And/or, to what extent is the organization effective because of the "brute force" of a big checkbook? Metaphorically, are we opening the door to success by easily turning the key in the lock, or are we battering down the door? Carefully focused research will be required to answer this question.

CHOOSING A MILITARY POLICY THE HARD WAY

Introduction

How does a nation truly know if it has an effective military policy? The only reliable way to answer that question, currently, is to go to war. That may be considered a form of empirical study and therefore an effective test. That test, however, is abhorrent in its human, environmental, and financial cost. In this chapter, I will test military policies in a way that is much less expensive; and yet, may be just as effective.

In casual conversation, and in many academic works, the term "military policy" is synonymous with "military action." For example, the question, "What should the military do in the present situation?" is often replaced with the question, "What should our military policy be in the present situation?" In the present chapter, I eschew that naive view and look at a military policy as representing a range of options within a military context (von Clausewitz, 2008, original, 1832). In this way, a military policy may be understood as a theory or lens through which the user may see and understand the world. A better military policy is one where the user may make better, more effective, and more useful decisions.

Military policy, as policy, is an understudied topic. In the voluminous "Encyclopedia of Policy Studies" (Nagel, 1994a) there is nothing specifically about military policy. In works where military policy is studied, military success (and defeat) is typically explored in a narrative and historical fashion. For example, in his well-cited history of US military policy Weigley (1973) writes primarily in a descriptive/ historical style.

For thousands of years, strategic thinkers have been searching for systematic understanding of war (Martel, 2007). From ancient times, military battles "were decided on strength and will" (Kahn, 2006: 125). Less attention was paid to more complex considerations of intelligence and strategy. In his classic work, "On War," Carl von Clausewitz (2008) describes the early evolution of military policy. First, he says, there was simple practice. Without much reflection, armies engaged in training, building forts, marching, and organization. As sieges became more common, some leaders developed tactics to overcome the stalemate of the siege. Those tactics were reflected in their written memoirs. However, those memoirs were not systematic. Therefore, they were difficult to understand and difficult to learn.

By Napoleon's conquest of Prussia in 1806, theories of warfare were often speculative including, "irreconcilable opposition… between theory and practice" (von Clausewitz, 2008: 70). Some theorists found focus by concentrating on the more easily quantifiable aspects of war. They would look at the size of armies and the number of cannon. Concepts such as the leadership, the potential actions of opposing armies, and the motivation of troops were not well understood by authors and so were left undefined (von Clausewitz, 2008).

Even today, the seemingly simple question of what constitutes "victory" in a military policy is thrown into doubt. Indeed, "Without a systematic answer to the question of what it means to be victorious, scholars and policymakers cannot judge what interests are at stake, whether it is prudent for the state to go to war, and what resources the state should commit to war" (Martel, 2007: 3-4). Thus, the present chapter may be understood as a very new approach to answering very old questions: How effective is our military policy; and, is our military policy better than that of our opponent?

While you (as reader and thinker) may not be a military strategist, you may find this comparison interesting for at least two reasons. First, this chapter may serve as a general example of how key measures of Complexity and Robustness provide useful insight into policy models— and provide you with the ability to more accurately choose between conflicting models (without the difficulty associated with actually applying the model—especially when that means engaging in warfare). Second, because the military approach may also be understood as a metaphor. In some conflicts, businesses battle over market share while in others individuals fight against injustice. Thus, by understanding the comparative advantage gained by adopting better policies, it becomes possible to more cogently choose our policies (whether one is in a corporate, NGO, or governmental organization) and so become more successful in our organizational pursuits.

In some of those pursuits, by way of example, we might expect to see corporate departments and social change organizations using claims of success to bolster their requests for additional funding. However, the hurly-burley of the market makes it very difficult to clearly determine the winner because corporate policies are replete with claims of success and counter claims of failure. So, here too, military policy provides a useful example because each military policy has a very specific life span and that life is bracketed by clear indicators of success or failure.

Another reason to study military policy is because a military may absorb a large share of national resources (even in peacetime). That means resources must be taken from some other allocation (e.g., social or economic development). Therefore, having a better understanding of military policy will enable a nation to distribute funds efficiently within the military as well as more efficiently between the military and other organizations.

Comparable cases are found in the Franco-Prussian war (1870-1871) because the combatants possessed armies of similar size. By analyzing their policies, it is possible to develop a new and deeper understanding of how a better military policy enables from effective performance.

Analysis

The Prussians, defeated by Napoleon in 1806, undertook to develop better military policy (Foch, 1903: 27, 323-326). When France attacked Prussia in 1870, the result was a decisive and surprising victory for Prussia. The two nations were of similar size and strength—but applied very different policies. What accounts for this turning of the tides? How do nations of similar size, wealth, and technological capacity gain the upper hand in war? While a complete analysis is beyond the scope of the present chapter, the present method adds a new perspective thus improving our understanding of how victory and defeat occur from a military policy perspective.

The French policy is drawn from Foch (1903) who describes the French military policy as it existed before the war. Foch notes that his is a "theory of war" or a "doctrine" (p. 7). Therefore, it may also be understood as a policy—a lens through which one might understand war. And, through that understanding, be able to take effective action.

The Prussian policy is based on the work of von Clausewitz who served as head of the Prussian military school. Further, his book was (and remains to this day) a classic in military policy. Therefore, it seems reasonable to accept his position as a valid representation of Prussian military policy of that time. He does not seem to sum up his theory in a very concise manner. So, I will rely on others—specifically Herwig (1998).

Dependent Variable—Success or Failure in War

France declared war against Prussia in 1870. They proceeded to lose that war, much to their surprise, when the Prussians defeated the emperor Napoleon III and his army rather rapidly. When the French declared themselves a new republic, and raised another army, the Prussians went on to defeat that one as well. In some sense, we might say that the Prussians won the war twice.

Complexity and Robustness

Policy #1—the Prussian model

Herwig, (1998: 68) states that the Prussian model of military policy includes communications, logistics, law, medicine, geography, tactics, history, weapons, fortifications, and staff work. Understanding these combined areas of study is needed to create an officer who is more likely to be successful in war. Therefore, this perspective may be represented graphically as Shown in Figure 3-1:

Counting the number of aspects I find a total of eleven within the model. Therefore, the Complexity of the Prussian military model may be understood as C=11. Only one of these aspects is concatenated (Opportunity for Victory). Therefore, the Robustness of the model is R=0.09 (the result of eleven total aspects divided by one concatenated aspect).

Von Clausewitz notes that, "war is a constant state of reciprocal action, the effects of which are mutual" (von Clausewitz, 2008: 71) suggesting that he had some understanding of Complexity and nonlinear thinking. Although these co-casual insights are not reflected in the

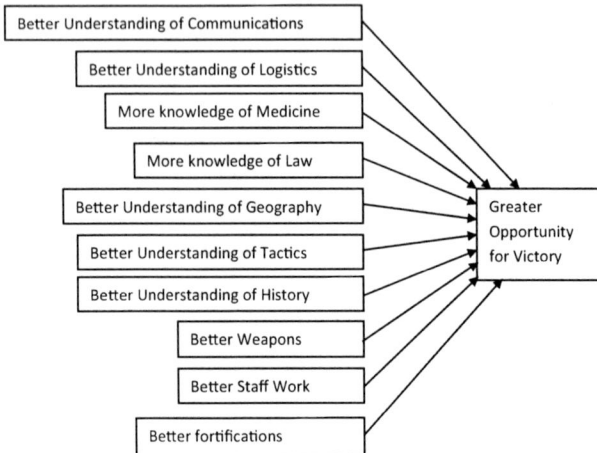

Figure 3-1 *Prussian Military Policy*

Prussian model, a PA analysis of his complete book may reveal a higher level of Robustness.

Policy #2—The French model

Foch (1903: 3) characterized the French theory of war prior to 1870 as focused on having (or seeking) More troops, Better rifles, Better artillery, and Better positions. Combining these, they believed, would ensure victory.

This can be diagramed as shown in Figure 3-2:

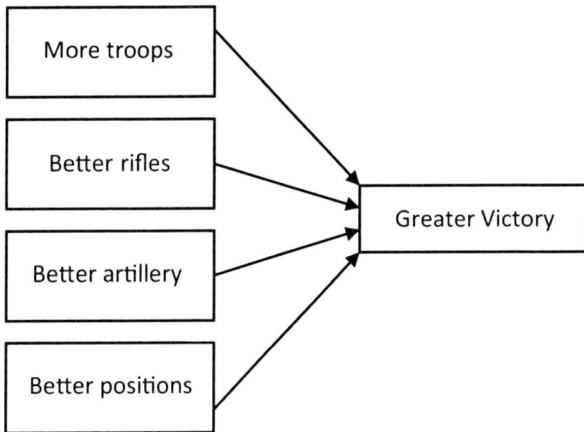

Figure 3-2 *French military policy*

Here, there are five aspects. Therefore, we may say that this policy of war has a Complexity of C=5. One of them (Victory) is concatenated. Therefore, this model has a Robustness of R=0.20 (the result of one concatenated aspect divided by five total aspects).

Comparison and Conversation

If all nations go to war with the expectations of success, and half of them are wrong, we must wonder how can a decision that is so important be made incorrectly so often? In this section, I will discuss similarities and differences between the French and Prussian military policies—how they were implemented and what the implication are for understanding and improving policies.

Complexity and Robustness

As seen in the above analyses, the Prussian policy (C=11) is about twice as complex as the French policy (C=5). An important point for the present section is that the difference in Complexity reflects a depth of understanding. Apparently, the difference found here made it relatively easy for the Prussians to secure a rapid victory with comparatively few losses.

Such a depth of understanding runs counter to the idea of "parsimony" in theory construction (e.g., McIntosh, 2007). It also provides another valuable lesson for practitioners suggesting that we should strive to learn as many perspectives as possible—simplicity may be comforting, but it is not a reliable ally.

Both models present a very low level of Robustness— although the French R=0.20 is about twice the Prussian R=0.09. The difference in Robustness may be influenced by the source of the data. Certainly, from his own description, von Clausewitz meant for all his methods to be interconnected—perhaps co-causal. This suggests the opportunity for a deeper study of his work that may reveal a higher level of functional Robustness.

It should be noted that the Robustness of the Prussian model (R=0.09) is less than the Robustness of the French (R=0.20). However (as with the low robustness of the UN and League of Nations constitutions), at this low end of the scale it may be that such a difference seems to be overshadowed by the vast gap in Complexity. More studies will be needed to clarify this relationship.

Resources and Implementation

The Prussian alliance included a population of 30 million people while the French nation contained a population of 38 million (Wawro, 2003: 19). Thus, from a population perspective, the French had an edge, although the Prussians were able to field more troops. Here, again, military policy played a vital role. The Prussian model called for universal conscription and rigorous training for the reserves. In contrast, the French levée en masse primarily recruited civilians to serve as soldiers for a short time.

Other resources were certainly important to the Prussian victory particularly, the command staff, advanced cannon, and well-trained reserves (Gray, 1992). Of these, the general staff appears to be closely related to the effectiveness of the Prussian model. Good policy and good governance go hand in hand as suggested in the previous chapter. Second, the effectiveness of the reserves may also be seen as closely related to the Prussian policy because that policy involved continues education. In comparison, the French reserves were mainly civilians who received minimal training.

Finally, the Prussian cannon were much improved over those of the French. This is an interesting point because both the Prussian and the French policies note the importance of artillery. It seems, however, that the Prussians followed their model more closely than the French. This, in turn, may suggest that the more Complex Prussian model provided the foundation for a richer, self-reinforcing, culture. In contrast, the French model provided more of a loose set of guidelines. Deeper discussion on the relative consequence of other comparable aspects of the two policies will require more space than is available in the present book.

Evolution of Policy

"Victory breeds arrogance; defeat drives reform" (Herwig, 1998: 72). Previously, defeats were understood to cause changes in military policy—but those changes have been understood in a more qualitative way. For a good example see "The Aftermath of defeat: societies, armed forces, and the challenge of recovery" (Andreopoulos & Selesky, 1994). For a view of evolving American military policy through a historical perspective, a good example is found in: "Arms and Men: A Study in American Military History" (Millis, 1981).

This perspective on change has strong parallels with our understanding of evolution theory as applied to organizations (e.g., Allen, 2003; Brown & Eisenhardt, 1997; Hull, 1988). Similarly, the concept of an evolutionary landscape suggests that increasing success creates a more narrow perspective (e.g., McCarthy, 2004; Rhodes & Donnelly-Cox, 2009). This has implications for gaining new insights into the de-evolution of policy—or why militaries are typically prepared to fight the previous war instead of the next one.

There may also be useful lessons to be found that might support a better understanding of the evolution of the social sciences in general—thus accelerating our ability to understand economic, psychological, and social issues. For example, apparently, the French were not able to learn the lessons from their first defeat in time to create a new policy that could be applied effectively. A few decades of effort, in contrast, enabled their creation of a policy with greater complexity and greater effectiveness. By identifying and quantifying these "trajectories of improvement" for a variety of policies, we can see what direction they are moving "forward." Understanding that direction enables us to improve those policies more easily and more rapidly.

Conclusions

The present analysis suggests that the Prussian military policy was more successful because it was more complex than the policy of the French. Both policies had a low level of Robustness it is difficult to draw a meaningful comparison on that score. It is also worth noting that the French did not adhere closely to their own policy—further weakening any potential benefit that they might have otherwise gained from its use. The present chapter also opens a new direction in the study of military policy. The innovative methods developed here for analyzing policy, as policy, suggest an emerging ability to understand our policies—and to make more effective policies that will enable us to take more effective action.

How do we know (as generals or analysts) if our policy is complex enough? The French believed that their military policy would allow them to easily win. How many other nations (organizations, corporations, etc.) hold the same mistaken belief? Clearly, intuition and self-confidence are not sufficient to answer this question. This is particularly important when our inability to answer such questions will cost enormous amounts of money, and determine the fate of individuals, corporations, NGOs, and nations.

In this analysis, there are also implications related to popular authorship and academic scholarship. First, the French believed they had the resources necessary for success. That view, again, is related to the French military policy. However, the French leaders could not comprehend the efficacy of the Prussian model. I suggest that the reason for this lack of understanding is because the French were viewing a complex (Prussian) model through the lens of a relatively simple (French) model. Thus, the relative complexity inhibited an improved understanding. This same problem exists today in many variations.

Consider, for example, how one firm may become quite successful because it has a highly Complex corporate culture. The Complexity of that culture (developed over decades of dedicated effort) enables it to better understand the market and out-perform its competitors. Because of that success, popular leadership books are written and scholarly analyses are undertaken. Those books and studies are supposed to help others learn from, and emulate, the original firm's success. The problem occurs when, in the process of research, analysis, writing, and publication, the understanding of that firm becomes "dumbed down." The readers, whether popular or scholarly, find themselves attempting to understand a complex situation from a simple perspective—a recipe for failure.

Ashby's law of requisite complexity may be invoked for the present study. For Ashby, "The law holds that for a biological or social entity to be adaptive, the variety of its internal order must match the variety imposed by environmental constraints" (Boisot & McKelvey, 2010: 421). Here, however, I am not talking about some fuzzy sense of Ashby's law—that our plans should be (or need only be) as complex as the environment. Particularly because, in most or all situations, neither the plan nor the environment are completely measurable. Rather, here there are two armies (each with their own policy), each of which is part of the environment of the other. Here, there appears to be something more akin to "dominant complexity." In short, the present chapter shows that policies that are more Complex and more Robust will tend to win over policies that are less so.

Finally, in peacetime as well as war, the cost of a military is staggering. One example of the importance of this field of study may be seen in the US military which required an expenditure of $651 billion in 2009 (http://www.gpoaccess.gov/usbudget/fy09/pdf/budget/defense.pdf). If methods developed in the present chapter can be used to improve the efficiency of our military by even one percent, we can have a military that is just as effective as it is today while saving six and a half Billion dollars each year. This, to any educated individual, suggests the value of pursuing a research program in this nascent field of critical metapolicy analysis.

Such a research program would also benefit by analyzing related policies (e.g., weapons research and technologies, economics, ethics, politics). And, the opportunity to rigorously integrate policies from differing fields suggests an intriguing opportunity for both research and practice.

WHEN GOOD EFFORTS CAUSE POOR RESULTS: HOW TO IDENTIFY PROBLEMATIC ECONOMIC POLICIES BEFORE THEY FAIL

Introduction

Economic policy is related to questions contrasting central planning and market control, socialist and capitalist perspectives, economic growth, deciding whether to allocate funds to military spending or public welfare and others (Albritton, 1994: 167-169). Issues of economics include unemployment, inflation, business, consumers, and labor (Nagel, 1994b: 880). Economics is not only about solving problems, it is also a tool to help us reach our loftiest goals, including zero unemployment and zero inflation (Nagel, 1994b: 900-902).

Approaches for coping with economic problems have existed since ancient times. These, however, might be termed "pre-theoretical" because they did not explicitly present the existence of a deep understanding of the economic situation. In the place of a deeper theoretical understanding, some ancient writings include references to supernatural causes. For example, the ability to be victorious over one's enemies is due to the "gracious favor of the gods" and difficult economic issues are due to an "unhappy star" (Schuettinger & Butler, 1979: 155). Does such a perspective count as a theory? For now, that must remain an area for future investigation.

While deities and astrological events are no longer central, our present understanding of economics is clearly insufficient to the task dealing effectively with the complex issues of our 21st century world (e.g., Colander *et al.*, 2009; Kates, 2010). Dobusch & Kapeller (2009) note that, "Economics is locked into neoclassical thinking, which prevents the emergence of possible alternatives regardless of their qualities and strengths" thus inhibiting the evolution and improvement of economic theory. At least one critic suggests that our entire economic paradigm is flawed because there are, "multiple views of social reality and policy problems and no definitive way to adjucate among them" (White, 1994: 862).

Indeed, neither micro economics nor macro economics has produced theories that can be used in predicting our economic future (Auyang, 1998: 368). Instead, we have models that are "woefully unrealistic" (Auyang, 1998: 329) because economic theories are founded on assumptions of rationality, stability, full employment, infinite flexibility, and perfect knowledge (Auyang, 1998: 206, etc.). Instead of effective theory and policy, "the hills are alive with the sound of new tools and jargon" (Freeman, 1998: 19) while various economic models are vying for domination in some "war of the models" (Freeman, 1998) with no clear winner in sight

Without a method for advocating rationally between policies, scholars may fall back on cultural or religious standards for choosing which policies to apply (Simons & Elkins, 2004). That kind of approach is similar to "coherence" where we determine the validity of a new based on how well it fits with existing theory (Kaplan, 1964: 311-322). If we rely on the concept of coherence for validating our economic policies, we might as well throw out the science of economics and return to reading religious texts for economic inspiration—a completely unsatisfactory approach.

One reason why it is difficult to develop effective policy is because social relations are "complex and multi-causal" (White, 1994: 862). One way to overcome that difficulty is using a complexity perspective. For one such example, Wollmershäuser (2003) compared computer models with historical data. He found that relatively simple policy rules, applied with the assumption of a closed economic system, would not perform as well as more complex rules working under the assumption of an open system. Essentially, the use of simple policies is associated with increased risk of failure when confronted with uncertainty. Readers are invited to make their own conclusions about how frequently we face uncertainty in our economic analyses.

In the present chapter, I use PA to analyze two economic policies. The Complexity and Robustness from those analyses will be compared with the historical results of implementing those policies. The goal is to provide new insights into the creation, analysis, and implementation of economic policies that may also be generalized to policies of all types. These insights, on our emerging ability to compare policies will enable policy analysts to make better policies. They will also make it easier for decision makers to choose between competing policies. Finally, with these insights, we may anticipate new opportunities for scholarly research.

The comparative case study approach calls for two cases with similar independent variables and dissimilar dependent variables (Lijphart, 1975). Here, the clear independent variable will be the economic policies of

Australia and the Netherlands. For the dependent variable, we will look at the results of implementing those policies including changes in inflation, GDP, and unemployment.

Analysis

These policies are comparable because both were created and implemented during the 1980s by industrial nations—each with a population of about 14—15 million people (United Nations, 1982). Both policies were created through a process of conversation and negotiation to alleviate a problem of national concern. One difference is that the Netherlands agreement was derived from a meeting between business and labor. While, in contrast, the Australian approach was primarily derived from a meeting between the government and labor. It is not immediately clear the extent to which the difference between the two may have had on their respective outcomes.

Additionally, it may be noted that these are comparable policies because they were both developed and implemented on a national scale. Both nations were democratic and industrialized. They both existed in the same global economy at the same time in history and during the same time of great economic upheaval and global changes in economic policies (Rodrik, 1996).

Independent Variable—Two Economic Policies

The first policy is the Wassenaar Accord (van Ours, 2002). In 1982, the Netherlands faced, "a severe recession, spiraling unemployment, and significant inflation" (Hemerijck & Vail, 2006). In order to boost employment and economic prosperity, negotiations were convened which included representatives from labor unions, employers' associations and government. The resulting one page document focused on a few key ideas (Hemerijck *et al.*, 2006).

• Recovery of economic growth will lead to structured improvement of employment.
• Stable price levels will cause structured improvement of employment.
• Improved competitiveness of enterprises will support structured improvement of employment.
• Improved competitiveness will also lead to improved profits for corporations.

Following this policy, the Netherlands began redistributing existing employment (including work-time reduction, part time work, reducing the unemployment of youth). These were undertaken with no increase in costs to employers.

The second policy is the Prices and Incomes Accord. Developed in 1983, in Australia, this agreement was to set the stage for wage and price controls to stabilize and improve the Australian economy. According to then Prime Minster Hawke (Hawke, 1983), A National Economic Summit Conference had been convened to address the "serious economic problems facing the Australian economy." The goals of this conference were to "promote employment," "achieve recovery and growth," avoid a "wages-prices spiral," develop better understanding of economic conditions, develop better approaches for tracking prices and setting wage/price levels, and "To secure broad agreement on the role of an incomes and prices policy."

Based on the Prime Minister's report, the "key elements" of this approach include:

- Serious economic problems require cooperative and innovative responses.
- Recovery requires restraint in expectations and claims (referring to wage and price increases).
- Maintenance of real wages is important.
- Centralized wage fixation is needed (not piecemeal).
- Price fixation is needed.
- More spending will require more taxation.
- More support is needed for individuals in need.
- More amalgamation of unions.
- Protection of rights—particularly women and children is important.
- Need for health care.
- Sharing information is important.

Dependent Variable—Success or Failure

For this analysis, I chose a set of key economic indicators and identified changes that occurred over a ten-year period. The indicators (GDP, unemployment, inflation rates) were used because they seemed to be good representatives of the policy goals—primarily employment, stability, and prosperity. A ten-year period was chosen because it seemed like a reasonable window where we might expect to see economic changes directly related to the policies.

The Wassenaar agreement set the stage for the "Dutch miracle" and since attained "a mythical status" (Welters & Muysken, 2004). Shortly after the accord was reached, the

Netherlands unemployment rate began a steady decline. The effect of the accords on inflation is less clear. Inflation was about 4% and declined immediately. Within five years, inflation was below zero and began to rise again. During the entire ten years, the GDP of the Netherlands enjoyed a steady climb. After the implementation of the Australian Prices and Incomes Accord their GDP also climbed steadily (Shane, 2010).

For the Australians, inflation was a bit of a wild ride— starting at about 11%, dropping to 4% then rising again to 9% before leveling at about 8% for a few years then dropping again. Unemployment in Australia likewise responded unpredictably to the accords. After an initial decline, unemployment later rose to a higher rate than before. Not exactly the stabilization that the government had been expecting!

Looking at the changes in the rates of inflation, for example, the Dutch inflation rate oscillated between zero and four percent, while the Australian inflation ranged from one percent to eleven percent. It seems reasonable to say, therefore, that the rate of Australian inflation was much less stable over time than was the Dutch rate of inflation. These are shown in Figure 4-1 and 4-2.

Other comparisons are possible. And, for deeper analyses, more indicators would be desirable. For the present study, these examples should prove sufficient.

Overall, the implementation of the Netherlands showed more improvement and more stability than did the implementation of the Australian policy. That is to say that the Dutch policy provided a more predictable path

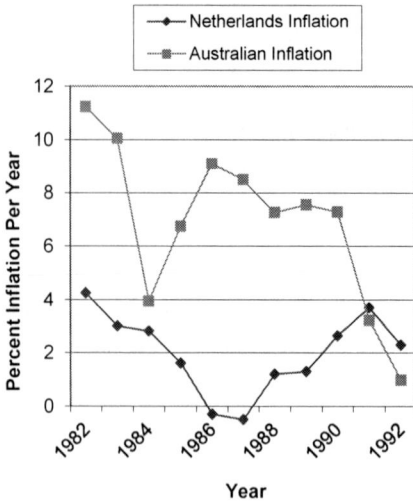

Figure 4-1 *Comparing Rates of Inflation of Netherlands and Australia (Inflation.eu, 2011; RateInflation, 2011).*

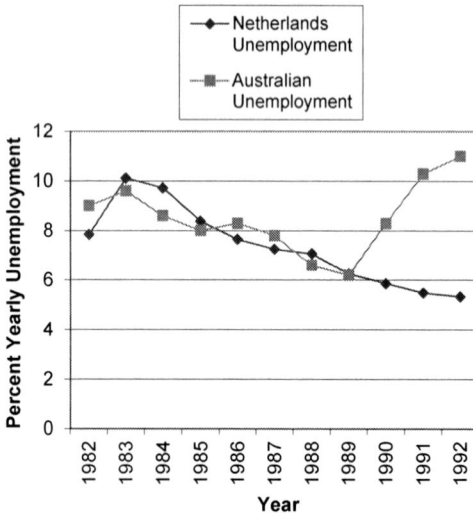

Figure 4-2 *Comparing Unemployment Rates of Netherlands and Australia (IndexMundi, 2011; Loundes, 1997)*

to achieving their desired goals. Some key differences in these policies and their results are shown in Table 4-1.

		Australia	Netherlands
GDP	**Effective?**	Yes	Yes
	Stability?	Good	Good
Unemployment	**Effective?**	Yes	Yes
	Stability?	Poor	Fair
Inflation	**Effective?**	No	Yes
	Stability?	Fair	Good

Table 4-1 *Comparative Usefulness and Stability of Economic Policies*

These results suggest the usefulness of measuring the Complexity and Robustness of policy so that they may be used to determine the potential effectiveness of economic policy in achieving policy goals. The stability suggests that the policy is effective at moving toward those goals in a predictable way, rather than a haphazard set of results. Where the results are not stable, it may be suggested that the policy may have reached its goal by good luck, rather than good policy planning.

Complexity and Robustness

Policy #1—The Wassenaar Accord

There are five aspects to this policy, therefore the Complexity of the policy may be understood as C=5. Of those five, one is concatenated (Structured Improvement of Employment), therefore the Robustness of the policy may be understood as R=0.20 (the result of one divided by five).

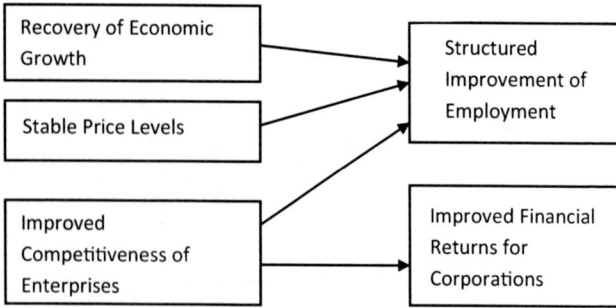

Figure 4-3 *Diagram of the Wassenaar Accord*

Policy #2—Prices and Incomes Accord

This policy contains a large number of concepts. However, many are not presented as causal statements. Instead, the concepts were presented as "important" or "needed." Graphically, this may be represented as shown in Figure 4-4:

Here, it can be seen that there are 14 aspects. Therefore, the Complexity of this policy is C=14. There are, however, no concatenated aspects. Therefore the Robustness of this policy is R=0. This will be discussed in the subsection below.

Comparisons and Conversations

Complexity and Robustness

The Australian plan is clearly the more complex of the two at C=14. That is nearly three times the Netherlands' C=5. Yet, when we look at the Robustness of the two policies, the Australian policy is zero compared to the Netherlands at R=0.20. Therefore, the greater Robustness of the highly successful Netherlands policy seems to have "trumped"

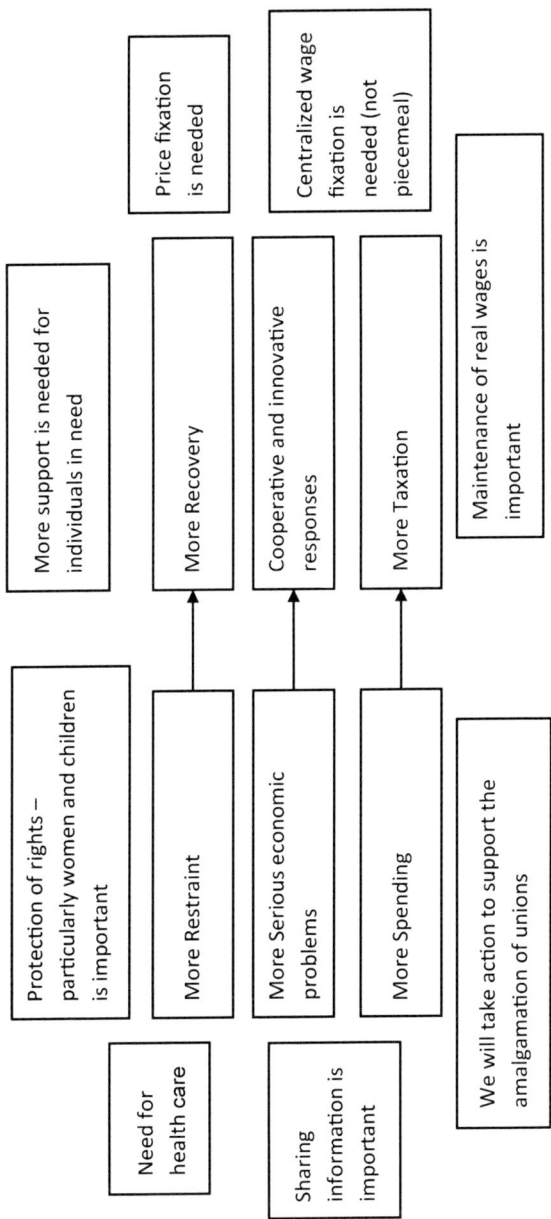

Figure 4-4 *Diagram of Prices and Incomes Accord*

Need for health care

Protection of rights – particularly women and children is important

More support is needed for individuals in need

Price fixation is needed

More Restraint → More Recovery

Sharing information is important

More Serious economic problems → Cooperative and innovative responses

Centralized wage fixation is needed (not piecemeal)

More Spending → More Taxation

We will take action to support the amalgamation of unions

Maintenance of real wages is important

the more complex but not so effective policy of Australia. This difference may be understood by considering the differences in the logical / causal connections between the concepts in the Australian policy (Figure 4-4).

There, the bulk of the aspects are not casually connected. They are, in essence, "truth claims"—essentially stating that something is true or "important" or "needed." Key here is the idea that those claims are unsupported. There is little or no discussion about why these things are important. Or, of greater significance to the structure of the policy, there is nothing to show what aspects might be causal to those important things. Or, similarly, what consequences might result from changes in those important things.

That is not to say, of course, that those aspects are (in some absolute sense) unimportant. Rather that the un-examined nature of those statements—of their importance— suggests that they were not truly understood. In short, this model contains many pronouncements but little understanding. Those disconnected pronouncements may be understood as conceptually similar to the idea of a protective belt of a theory while the casually connected concepts may be understood as the core (Wallis, 2008a, 2009a). The belt protects the core from challenges and criticism (Lakatos, 1970); a stance that is more important for political spin than for the creation of useful policy.

The other aspects of the less-successful Australian policy in Figure 4-4 are seen as "linear logics." That is to say they are understood as simplistic cause-and-effect mechanisms. Without the depth of understanding evidenced by concatenated propositions, it becomes very difficult to understand the situation and find a solution to economic problems.

As noted above, the goal of the Wassenaar accords was to boost employment and economic prosperity. In this, they were generally successful. The goals for the Australian Prices and Incomes Accord were to stabilize and improve the Australian economy, promote employment, and achieve recovery and growth. This, they did not do so well—particularly in their quest for stability. Given those goals, and the results presented above, it seems that the Wassenaar Accords (R=0.20) was a much more effective policy than the Price and Income Accords (R=0). Even though the Australian policy was more complex (C=14) than the Dutch (C=5).

Conclusions

In this chapter I used propositional analysis to investigate two economic policies. The Complexity and Robustness of those policies were compared with changes in key economic indicators for a period of approximately ten years. Policies of greater Robustness (as measured by the internal integrity of the logical structure) appeared to be more useful for achieving economic goals. Also, it seems that more Robust policies will help us reach those goals with greater stability (avoiding the roller-coaster ride of unpredictable economic change).

Boettke (1997: 13) discusses two approaches to deciding between economic theories. One relates to the internal coherence of the theory while the other relates to the empirical correspondence between the theory and the outside world. Where he suggests, "economists must steer a course between [the two]" the present chapter suggests that the two approaches must be rigorously integrated. The analytical process that Boettke derides as "self-indulgent mental gymnastics" must be replaced with rigorous and repeatable analytical methods such as propositional analysis and cliometric metatheory. The approach presented in this chapter gives us the opportunity to advance economic theory more rapidly. Such an advance is necessary and beneficial to our global and national economies—and to the economic life of every person on our planet.

THE FUTURE OF POLICY

Introduction

I t is not easy to develop effective policy that is highly useful for understanding the world, predicting events, guiding decision making, and helping communities and organizations to achieve their goals. Indeed, it is rather like assembling a jigsaw puzzle in the dark. Despite the obvious difficulties, the present analysis has provided a feel for how some of the pieces might fit together. The present book has advanced the field of policy studies by identifying previously unrecognized indicators that are understood as important because of the causal relationships between them—how they fit together. On one level, this has been about how casual relationships exist within the propositions found in policy texts; on another level, this has been about how the implementation of those policies may be judged effective.

To date, the process of evaluating policy prior to implementation has been primarily based on intuition informed by education and experience. Those evaluations, particularly in the public sphere, may be overshadowed by partisan conflict. Previously, there has been no rigorous measure of internal coherence so it has been difficult (or impossible) to know in advance if the policy is likely to fail. This has been a troubling issue in many policy fields including economics, military, and International Relations.

It has been similarly difficult to evaluate the effectiveness of policies during and after implementation. Within some fields (such as military science), there is some clarity. Even with that clarity, however, policies still fail about half the time. In other fields (such as economics), it is difficult to know whether a policy has succeeded or not. Certainly, we have room for improvement. And, by improving our policies, we stand to make tremendous gains.

The complexity theory based PA method, tested and advocated in this book, suggests that we now have an effective and (to some extent) more objective tool for evaluating policies prior to implementation. And, critically, that tool can be used to predict the success or failure of a policy prior to implementation. In this chapter, I will present some limitations of the present analysis, some opportunities for future studies along with some implications for scholars, practitioners, and metatheoreticians.

Limitations

The present study does, of course, have room for improvement. The space allocated for the present book does not allow for a complete analysis. And, more relevant to the present conversation, today's understanding of policy and theory cannot be relied upon to tell us what the "important" things are that we "should" be analyzing.

A notable limitation of the present study is the lack of highly Robust policies to analyze. The policies studied here have all been at the low end of the Robustness scale—between R=0 and R=0.20. Without such policies, we cannot compare highly Robust policies with policies of lower levels of Robustness. A hypothetical example of such a comparison (one policy with R=.08, the other R=0.67) has been explored in some depth. That study suggests a variety of benefits including reduced inter-agency conflict, greater community collaboration, and improved ability to achieve policy goals (Wallis, 2010d). More studies are needed to find and compare such policies.

Another limitation of this study is the difficulty addressing aspects of policy-making that are outside the formal model or "off the page," such as the mood, health, superstitions, or prejudices that might alter a policy practitioner's decision-making process. Additionally, this study is limited because it compares only two models in each case study. Future studies should compare more models, models from a wider range of policy areas, and compare policies with the results of their implementations. For such an approach, I strongly suggest the cliometric metatheory approach (e.g., Faust, 2005; Faust & Meehl, 2002; Meehl, 1992, 2002, 2004).

Future Studies

The PA method has been shown to be useful in identifying policies that will support more predictable change. An excellent opportunity exists to use this method to measure emergent systems and related non-predictable events in at least two ways. First, within the application of a single policy, we might study those parts of the policy where events and change are understood and manipulated. While, in contrast, those parts of the policy which are not so well understood may be those where we find the most emergence and unpredictable change. Second, considering a broader context, it would be interesting to study policy implementation and the unanticipated consequences that occur outside the specific realm of that policy. For example, consider the expensive "war on drugs" draws resources away from other (seemingly stable) community systems. That change may destabilize those communities, thus causing more problems (one of which may be increased drug use). Another area for study might compare areas where policy is applied with areas where no formal policy exists.

In general, most analyses in the present book included only representational sections of more complete policy text. The constitutions of IGOs were limited to the preambles, the military policies were focused on a set of key propositions, and the economic policies included a policy pronouncement by a self-interested politician. Future studies could improve on these analyses by conducting similar metapolicy analyses of the complete documents.

The opportunity exists to study the evolution of policies. As suggested in the evolution of military policy and the evolution of IGO constitutions, there appears to be an increase in the Complexity of those policies over time. This correlated with the evolution of theory over time as the theory moves from simple (and ineffective) to more complex (and more useful) (Wallis, 2010a). If the trend of policy evolution continues to match the trend of theory evolution, we can expect to see policies become increasingly complex over time. And, as they become more complex, we can expect them to become *slightly* more effective. Then, as they mature, they will become more Robust and *much more* effective.

This opens an important question of how much complexity is needed to reach "Complexity apogee?" That is to say, at what point do our policies become so large that we begin to crystallize the embedded understandings into policies that are less Complex, yet more Robust?

Implications for Practitioners

The analyses presented in this book suggest important and useful implications for practitioners in NGOs, corporations, governments, and others. In general, the key suggestion for this book is that practitioners (including citizens participating in conversations of the public sphere) should carefully analyze every policy for Complexity and Robustness. And, before implementing a policy, strive to increase both of those. Similarly, when faced with two competing policies, the best idea would be to choose the policy that has the highest level of Complexity and Robustness.

For creating more effective organizations, it may seem on the surface that one need merely create a larger (more complex) constitution. However, there are a few important caveats to consider. First, size is not Complexity: A constitution containing one hundred thousand words will not be better than one containing ten thousand words if there are the same number of concepts in each document. Second, Robustness counts. The interrelationships between those concepts are an important measure of the extent to which the document reflects the world. Finally, a constitution that is more complex will tend to require more people with greater skills, and more resources to enact that kind of constitution effectively. Note, this is different from simply needing "more resources," it is having more resources that are more carefully integrated, and have a more useful and effective understanding of the world.

Similarly, it is important to focus on the core of the policy. It is not wise to be distracted by pronouncements and truth claims (e.g., "health care is important"). Because, no matter what the emotional influence of such statements, their main strength lies in their ability to confuse the issue—not resolve it.

Implications for Scholars

This book opens new approaches for fields of constitutional studies, political science, International Relations, governmental studies, economic policy, economic theory, military policy, organizational theory and many others. Scholars may add a new and promising dimension to their research by using the methods presented here to analyze a variety of documents that include policies or are policy-like (or theory-like). As scholars find new insights into poorly understood successes and failures of organizations of all sizes, breakthrough advances in our ability to create effective policy seem probable.

For IR scholars, specifically, this new approach cuts the Gordian knot of the "great debates" by providing finely honed analytical tools to replace the bulky tangled ropes of realist, constructivist, and liberal metatheory (where metatheory is understood as a vaguely defined perspective rather than critical metatheory where specific tools are applied for analysis).

For scholars who are interested in studying the relationship between policy and success, the present book suggests that textual or historical exposition is not sufficient. Instead, we need to measure the Complexity and Robustness of each policy. We also need to quantify the relationship between the specific aspects of the policy and how those aspects were implemented and the results. What happens, for example, if one aspect of that policy is ignored? Will this be a bigger problem for highly complex policies or for simple policies?

There are countless policies that may be mined for their propositions and no lack of historical data to link with the implementation. For military policies, and others, there may be subsequent policies. Finding the Complexity and Robustness of those agreements is likely to reveal new and important insights into the nature of policy evaluation.

Implications for Metatheoreticians

This book suggests the benefits of having a policy that is more Complex. The benefit of Complexity is important in the field of metatheory because it disconfirms the idea of parsimony that has been suggested by numerous scholars from Ockham to the present (e.g., Edwards, 2010; McIntosh, 2007). Where the rule of parsimony suggests that the more concise policies would be more successful, the present research shows the opposite tends to be true. This result supports arguments that the standard of parsimony is problematic (Dubin, 1978; Meehl, 2002) and validates the assertion that, "More complex models are more effective"(Wallis, 2010d: 164).

Pragmatically, it seems reasonable that more complex policies, like multiple views, provide a richer, more useful understanding (e.g., Bolman & Deal, 1991; Edwards & Volkmann, 2008). The reason the more complex perspective is an improvement is because a simpler perspective (containing fewer aspects) means that its sensemaking ability will "catch" fewer areas of knowledge and understanding. Thus, as the theory sees less, it will tend to miss more; particularly, "easily unperceived effects" (François, 2008).

In addition to better understanding individual policies, there is a vast opportunity to create a rigorous process of integrating multiple polices. Outside of policy analysis, PA has been used for deconstructing multiple theories into their component propositions, and then recombining those propositions to create a new, rigorously integrated theory. By using PA as a tool for integrating multiple policies it may become possible to rigorously and critically integrate policies in a way that improves the Complexity and the Robustness of the policy at the same time. Thus, we may be able to more rapidly accelerate our ability to improve our policies.

One concept suggested by the present analysis is the unexamined relationship between theory and policy. That is to say, at what point does one inform the other? And, at what point might they be understood as the same thing—a lens for understanding the economic and political world around us. Another question for metatheoreticians may be understood in terms of a "scale of abstraction." That is to ask if it is appropriate in one policy model to combine solid actions (e.g., working harder to increase production) with more abstract concepts (e.g., fixing an interest rate)?

Finally, if two policies have different levels of Complexity and Robustness, yet are seen as similarly efficacious, this may imply that the Robustness of one may be approximately equivalent to the Complexity of the other. Thus, drawing on the example of economic policy in the previous chapter, C=14 might be approximately equal to R=0.20. More study is needed in this area. Such an understanding of our economic policies might suggest, for example, that a theory of R=0.60 might be equivalent to a theory of C=42. While that is a large number of aspects to keep in mind, that difficulty might be reduced by employing a close-knit team of economists.

Conclusion

Policy is important because it serves as a lens that enables us to understand and take effective action to reach our goals. Policies are implemented because decision-makers believe they will be effective. However, military policy clearly fails 50% of the time; while, for economics and other policies, we are often uncertain of failure or success. This puts policy practitioners uncomfortably close to gamblers and weather forecasters; did the policy work, or were we just lucky?

Despite the vast differences between policies of war, economics, and international relations, the present book has found a common thread in the analysis of their inherent logics. With this approach, PA appears to be a "practical metatheory" that provides clear insights on how to improve policy.

Clearly, previous methods for validating policy prior to implementation (e.g., intuition, cultural norms, religious norms, empirical tests, plausibility, assumptions, etc.) have not proved effective. It does not seem that we can point to any string of policy successes; or to any trends that show we are improving our ability to create effective policy. The overarching hypothesis presented at the start of this book asked if there is a correlation between the Complexity and Robustness of a policy document—and the effectiveness of that policy in practical application. Based on the comparative case studies presented, the answer is yes.

While the analyses in this book have been historical, it is important to note that these methods can, and should, be used to analyze policy in the present to decide what to implement in the future. As we develop more effective policies, concerns around unanticipated changes may require a policy around how policies are developed and implemented. This would be a general form of meta-policy that may be implemented in concert with the critical metapolicy approaches presented in this book.

Of course, the structure of policy is not the only measure. It is also important to consider data gathering, analysis, and how the policy is implemented. Improvements in each of these areas will add to the efficacy of policy. However, as those factors have already been addressed by other scholars, the present book focuses on following a new line of investigation into how we might improve the usefulness of our policies.

To the extent possible within the exposition of a single methodology in a small space, the methods presented in this book represent a revolutionary advance over previous methods of analyzing policy through a historical or narrative approach because it provides a rigorous and quantifiable understanding of the internal logics of the policy. This new method is also an improvement over testing policy through the only "empirical" approach of application. Such applications are particularly problematic when a nation is deciding to spend thousands of lives in declaring war or trillions of dollars attempting to alleviate a fiscal crisis.

For nations (and other organizations) that suffer from polarizing debates over policy, the methods developed here suggest an additional benefit. Specifically, PA provides an objective perspective on policy evaluation that may serve to reduce the conflict. PA could similarly serve to reduce the acrimony between wealthy and developing nations as they struggle to find a policy that will benefit world as a whole.

The comparative case studies (constitutional, military, and economic) presented in this book suggest that there is an evolution in progress as policies move towards greater Complexity, greater Robustness, and greater usefulness. And, because that process is subject to muddling, political wrangling, and other uncertainties, the evolutionary path is not direct, smooth, or certain. Therefore, we cannot be certain that each new policy is more useful— or merely more expedient. PA provides a rigorous and workable method for developing more effective policies and better predicting the usefulness of a policy. Those measures, if appropriately applied in academic, policy, and public spheres, will help us move forward with greater certainty, and greater success in creating more effective organizations, international treaties, stable economies, greater justice, and achieving our most noble goals.

GLOSSARY

Aspect

The part of policy that represents a concept, idea, or notion. The aspect may be as concrete as in "apple" or as abstract as in "truth." An aspect is typically detectable, that is to say empirically measurable, but that is not an absolute standard.

Atomistic Logic

A kind of logical structure found within a proposition that is reductionist such as "A is valid" or "A is true." Or, more concretely, "Apples are important."

Branching Logic

A logical structure found within causal propositions including three or more aspects where a change in one aspect causes change in two or more other aspects. For example, a branching proposition might say that changes in A will cause changes in B and C. For a more concrete example, "More teamwork will lead to more cohesion, *and* more results, *and* more frustration.

Complexity

A measure representing the number of aspects within a policy. The calculated diversity of ideas within a policy document. For an abstract example, consider a policy containing the propositions: A is true; More B causes more C; More B causes more D. In such a model, there are four aspects (A, B, C, D). Therefore, the Complexity of the policy is C=4.

Concatenated Logic

A logical structure found within a casual proposition including three or more aspects where changes in two or more aspects cause change in another aspect. For an abstract example, a concatenated proposition might state that changes in aspect A *and* aspect B will cause changes in aspect C. In that example, C is the concatenated aspect, while A and B exist within a concatenated relationship but are themselves not concatenated. For a more concrete example, "More collaboration *and* more shared goals will result in more teamwork." Here, "teamwork" it the concatenated (and better understood) aspect.

Causal Relationship

Where two or more aspects are related so that a change in one causes a change in one or more others. A causal relationship is often expressed as a proposition, hypothesis, or a diagram. A causal relationship such as, "More A causes more B." may also be used as a general term in place of other more specific terms. Instead of saying "more" other indicators might be "better" or "less" (for example). Similarly, instead of "causes" more specific indicators might include such terms as "creates," or "engenders." In any case, the description must be specific to be valid. It is not useful to state (for example) that "A and B are interrelated" or "More A may cause more B" because the nature of the relationship is not causally defined. Logical structures often describe casual relationships (e.g., linear, branching, concatenated)

Critical Metapolicy Analysis

Rigorous and repeatable investigation of policy document according to a carefully structured methodology to quantify some understanding of the policy.

Dimension (or scalar dimension)

An aspect of policy that represents quantitative or qualitative variations. For example, a dimension of "size" might be used to represent whether a system is smaller or larger.

Functional (Complexity / Robustness)

The idea of functionality suggests that a policy may have a different level of Complexity or Robustness than is clearly identifiable in the policy text. That is to say, the Complexity and Robustness have not been effectively communicated through the text so they are not adequately addressed in the analysis. There may be, for example, tacit assumptions that are in opposition to (or in concert with) the explicit policy. The usefulness may be higher or lower than indicated by the text.

Integrative Analysis

Combined processes of qualitative and quantitative analysis involving rigorous hermeneutic deconstruction of text and rigorous re-integration of multiple texts following a structured methodology.

Linear Logic

A logical structure found within a proposition describing simple causal relationship between two aspects. Such as, "More A causes more B." Both A and B exist in linear relationship to one another. Here, A is the causal aspect and B is the resultant aspect). Linear structures can be of any length (e.g., More A causes more B which causes more C which causes more D... which causes more Z). For a more concrete example, "Having more shared goals leads to more teamwork which, in turn, leads to more productivity." Within an explanation, this may also be phrased as, "A is true because of B and B is true because of C... because of Z).

Logic Model

A set of interrelated logic statements such as a theory or a Policy Model.

Metapolicy

Generally the study of policy. This may be rigorous as in the use of propositional analysis or fuzzier as in the use of historical narrative. May also be used to describe a policy on how to make policy.

Metatheory

Primarily the study of theory, including the development of overarching combinations of theory, as well as the development and application of theorems for analyses that reveal underlying assumptions about theory and theorizing.

Policy

A cognitive structure (like a theory) representing how a community or organization understands the world, thus enabling them to take specific actions to achieve their goals. The policy, as a cognitive sense-making structure, does not (strictly speaking) include goals or actions.

Policy Model

Concise representation of a relatively complete policy. May be explicit—as in a diagram or it may be relatively implicit as found in a set of causal propositions.

Proposition

"A proposition is a declarative sentence expressing a relationship among some terms." (Van de Ven, 2007: 117). For example, "More travel leads to more discovery." (See Causal Relationship).

Propositional Analysis

A process of metapolicy analysis for investigating a policy to determine the Complexity of the policy (diversity of concepts) and the Robustness of the policy (the ratio of concatenated aspects to the total number of aspects). For an abstract example, consider a theory which states that changes in A *and* changes in B cause changes in C. Such a theory contains three aspects (A, B, C), one of which is concatenated (C). Therefore, the Robustness of the theory is 0.33 (the result of one concatenated aspect divided by three total aspects).

Robustness

A ratio describing the interrelatedness between aspects of a policy on a scale of zero to one. Robustness is calculated by dividing the number of concatenated aspects by the total number of aspects in a policy (see Propositional Analysis). Robustness is a measure of how well integrated the propositions of a policy are, the degree to which they are understood as existing in a systemic relationship, and the level of co-causality between the aspects. Robustness is also related to the effectiveness of the policy in practical application.

Social Science

The purposeful study and advancement of understanding in all fields of human interaction including (but not limited to) psychology, sociology, policy, ethics, business, management, human development, organizational development, economics, and social anthropology.

Theory

An ordered set of assertions. Weick (1989: 517. Drawing on Southerland).

REFERENCES

Albritton, R. B. (1994). Comparing policies across nations and over time. In S. S. Nagel (Ed.), *Encyclopedia of Policy Studies*, 2 ed., ISBN 0824791428, pp. 159-176.

Allen, P. (2003). Understanding social and economic systems as evolutionary complex systems, paper presented at *Uncertainty and Surprise: Questions on Working with the Unexpected and Unknowable*, University of Texas, Austin.

Andreopoulos, G. J. and Selesky, H. E. (1994). *The Aftermath of Defeat: Societies, Armed Forces, and the Challenge of Recovery*, ISBN 9780300058536.

Annan, K. A. (2000). *"We the Peoples:" The Role of the United Nations in the 21st Century*, ISBN 9789211008449.

Appelbaum, R. P. (1970). *Theories of Social Change*, ISBN 841040192.

Arbaugh, J. B. (2008). From the Editor: Starting the long march to legitimacy. *Academy of Management Learning & Education*, ISSN 1537-260X, 7(1): 5-8.

Auyang, S. Y. (1998). *Foundations of Complex-system Theories in Economics, Evolutionary Biology, and Statistical Physics*, ISBN 0521778263.

Bastedo, M. N. (2005). Metapolicy: Institutional change and the rationalization of public higher education, paper presented at the *American Educational Research Association*, Montreal, Canada.

Bastedo, M. N. (2007). Sociological frameworks for higher education policy research. In P. J. Gumport (Ed.), *The Sociology of Higher Education: Contributions and Their Contexts*, ISBN 9780801886157, pp. 295-316.

Bateson, G. (1979). *Mind in Nature: A Necessary Unity*, ISBN 0525155902.

Baum, D. (1996). *Smoke and Mirrors: The war on Drugs and The Politics of Failure*, ISBN 0316084468.

Bentz, V. M. and Shapiro, J. J. (1998). *Mindful Inquiry in Social Research*, ISBN 0761904093.

Bernier, L. and Hafsi, T. (2007). The changing nature of public entrepreneurship. *Public Administration Review*, ISSN 0033-3352, 67(3): 488-503.

Boettke, P. J. (1997). Where did economics go wrong? Modern economics as a flight from reality. *Critical Review*, ISSN 0891-3811, 11(1): 11-64.

Boisot, M. and McKelvey, B. (2010). Integrating modernist and postmodernist perspectives on organizations: A complexity science bridge. *Academy of Management Review*, ISSN 0363-7425, 35(3): 415-433.

Bolman, L. G. and Deal, T. E. (1991). *Reframing Organizations: Artistry, Choice, and Leadership*, ISBN 9780787964269.

Boschken, H. L. (1994). Organizational performance and multiple constituencies. *Public Administration Review*, ISSN 0033-3352, 54(3): 308-312.

Boudon, R. (1986). *Theories of Social Change* (J. C. Whitehouse, Trans.), ISBN 0745609503.

Brown, S. L. and Eisenhardt, K. M. (1997). The art of continuous change: Linking complexity theory and time-paced evolution in relentlessly shifting organizations. *Administrative Science Quarterly*, ISSN 0001-8392, 42(1): 1-34.

Budget. (1930). "Budget of the League," http://indiana. edu/~league/pictorialsurvey/lonapspg30.htm

Burrell, G. (1997). *Pandemonium: Towards a Retro-Organizational Theory*, ISBN 0803977778.

Casey, D. and Brugha, C. M. (2005). Questioning cultural orthodoxy: Policy implications for Ireland as an innovative knowledge-based economy. *Emergence: Complexity & Organization*, ISSN 1521-3250, 7(1): 43-59.

Chernoff, F. (2002). Scientific realism as a meta-theory of international politics. *International Studies Quarterly*, ISSN 1468-2478, 46(2): 189-207.

Chittick, W. O. (2006). *American Foreign Policy: A Framework for Analysis*, ISBN 9781933116624.

Colander, D., Goldberg, M., Haas, A., Juselius, K., Kirman, A., Lux, T. and Sloth, B. (2009). The financial crisis and the systemic failure of the economics profession. *Critical Review*, ISSN 0891-3811, 21(2-3): 249-267.

Commons, M. L., Trudeau, E. J., Stein, S. A., Richards, F. A. and Krause, S. R. (1998). Hierarchical complexity of tasks shows the existence of developmental stages. *Developmental Review*, ISSN 0273-2297, 18(3): 238-278.

Courchene, T. J. (1999). Subnational budgetary and stabilization policies in Canada and Australia. In J. M. Poterba (Ed.), *Fiscal Institutions and Fiscal Performance*, ISBN 0-226-67623-4, pp. 301-348.

de Green, K. B. (Ed.). (1993). *A Systems-Based Approach to Policymaking*, ISBN 9780792393368.

Dekkers. (2008). Adapting organizations: The instance of Business process re-engineering. *Systems Research and Behavioral Science*, ISSN 1092-7026, 25(1).

deLeon, P. (1999). The stages approach to the policy process: What has it done? Where is it going? In P. A. Sabatier (Ed.), *Theories of the Policy Process*, Vol. 1, ISBN 9780813343594, pp. 19-32.

Dennard, L., Richardson, K. A. and Morçöl, G. (Eds.). (2008). *Complexity and policy analysis: Tools and concepts for designing robust policies in a complex world*, ISBN 9780981703220.

Der Derian, J. (1996). Hedley Bull and the idea of diplomatic culture. In R. Fawn and J. Larkins (Eds.), *International Society After the Cold War: Anarchy and Order Reconsidered*, ISBN 9780333659564, pp. 84-100.

Detomasi, D. A. (2007). The multinational corporation and global governance: Modelling global public policy networks. *Journal of Business Ethics*, ISSN 1573-0697, 71(3): 321-334.

Diehl, P. F. (Ed.). (2005). *The Politics of Global Governance: International Organizations in an Interdependent World*, ISBN 1588263282.

Dobusch, L. and Kapeller, J. (2009). "Why is economics not an evolutionary science?" New answers to Veblen's old question. *Journal of Economic Issues*, ISSN 0021-3624, 43(4): 867-898.

Dror, Y. (1994). Basic concepts in advanced policy studies. In S. S. Nagel (Ed.), *Encyclopedia of Policy Studies*, 2 ed., ISBN 0824791428, pp. 1-30.

Dubin, R. (1978). *Theory Building* (Revised ed.), ISBN 002907620X.

Edwards, M. (2010). *Organisational Transformation for Sustainability: An Integral Metatheory*, ISBN 9780415801737.

Edwards, M. and Volkmann, R. (2008). Integral theory into integral action, Part 8., *Integral Leadership Review*, ISSN 1554-0790. http://www.integralleadershipreview.com/archives/2008-01/2008-01-edwards-volckmann-part8.html

Elliott, E. and Kiel, L. D. (Eds.). (1999). *Nonlinear Dynamics, Complexity and Public Policy*, ISBN 9781560727071.

Eloranta, J. (2010). Why did the League of Nations fail? *Cliometrica*, ISSN 1863-2513, 5(1): 27-52.

Etzioni, A. (2010). Behavioral economics: A methodological note. *Journal of Economic Psychology*, ISSN 0167-4870, 31(1): 51-54.

Faust, D. (2005). Why Paul Meehl will revolutionize the philosophy of science and why it should matter to psychologists. *Journal of Clinical Psychology*, ISSN 0021-9762, 61(10): 1355-1366.

Faust, D. and Meehl, P. (2002). Using meta-scientific studies to clarify or resolve questions in the philosophy and history of science. *Philosophy of Science*, ISSN 0007-0882, 69: S185-S196.

Foch, F. (1903). *The Principles of War* (H. Belloc, Trans.) (1 ed.), ISBN 9780548769287.

François, C. (2008). Complexity, a challenge to governance - postscript from a friend. *Systems Research and Behavioral Science*, ISSN 1092-7026, 25(2): 355-357.

Freeman, R. B. (1998). War of the Models: Which Labour Market Institutions for the 21st Century?, *Labour Economics*, 5(1): 1-24, ISSN 0927-5371.

Freire, L. G. (2011). *Potentials and pitfalls of metatheory in IR.* Paper presented at the WISC Conference 2011, Porto, Portugal.

Ghoshal, S. (2005). Bad management theories are destroying good management practices. *Academy of Management Learning & Education*, ISSN 1537-260X, 4(1): 75-91.

Goodrich, L. M. (1947). From League of Nations to United Nations. *International Organization*, ISSN 0020-8183, 1(1): 3-21.

Gray, W. E. (1992). *Prussia and the Evolution of the Reserve Army: A Forgotten Lesson of History*. U. S. Army War College, Carlisle, PA.

Hawke, R. (1983). National economic summit conference - Ministerial Statement, *House of Representatives*: 90. Canberra: Government of Australia.

Hayden, F. G. (2005). *Policymaking for a good society: The Social Fabric Matrix Approach to Policy Analysis and Program Evaluation*, ISBN 9780387293691.

Hemerijck, A. C. and Vail, M. I. (2006). The forgotten center: The state as dynamic actor in corporate political economies. In J. D. Levy (Ed.), *The state after statism: New state activities in the age of globalization and liberalization*. ISBN 978-0674022775, pp. 57-92.

Hernandez, M. and Hodges, S. (2001). Theory-based accountability. In M. Hernandez and S. Hodges (Eds.), *Developing Outcome Strategies in Children's Mental Health*, ISBN 9781557665201, pp. 21-40.

Herwig, H. H. (1998). The Prussian model and military planning today. *Joint Force Quarterly*, ISSN 1070-0692, 67 (Spring): 67-75.

Hjørland, B. (2002). Domain analysis in information science: Eleven approaches - traditional as well as innovative. *Journal of Documentation*, ISSN 0022-0418, 58(4): 422-462.

Hood, W. W. and Wilson, C. S. (2002). Analysis of the fuzzy set literature using phrases. *Scientometrics*, ISSN 0138-9130, 54(1): 103-118.

Hoppe, R. (2002). Cultures of public policy problems. *Journal of Comparative Policy Analysis: Research and Practice*, ISSN 1572-5448, 4(3): 305-326.

Hull, D. L. (1988). *Science as a Process: An Evolutionary Account of the Social and Conceptual Development of Science*, ISBN 9780226360515.

IndexMundi. (2011). "Netherlands unemployment rate," http://www.indexmundi.com/netherlands/unemployment_rate.html

Inflation.eu. (2011). "The Netherlands," http://www.inflation.eu/inflation-rates/the-netherlands/historic-inflation/cpi-inflation-the-netherlands-1982.aspx

Jacobs, L. R. (1995). Politics of America's supply state: Health reform and technology. *Health Affairs*, ISSN 1544-5208, 14(2): 143-158.

John, P. (2003). Is there life after policy streams, advocacy coalitions, and punctuations: Using evolutionary theory to explain policy change? *Policy Studies Journal*, ISSN 0190-292x, 31(4): 481-498.

Kahn, D. (2006). The rise of intelligence. *Foreign Affairs*, ISSN 0015-7120, 85(5): 125-134.

Kaplan, A. (1964). *The Conduct of Inquiry: Methodology for Behavioral Science*, ISBN 0765804484.

Kates, S. (2010). *Macroeconomic Theory and Its Failings: Alternative Perspectives on the Global Financial Crisis*, ISBN 978-1848448193.

Kerr, D. H. (1976). The logic of 'policy' and successful policies. *Policy Sciences*, ISSN 0032-2687, 7: 351-363.

Kessler, E. H. (2001). The idols of organizational theory from Francis Bacon to the Dilbert Principle. *Journal of Management Inquiry*, ISSN 1056-4926, 10(4): 285-297.

Kingdon, J. W. (1997). *Agendas, Alternatives, and Public Policies* (2 ed.), ISBN 0673523896.

Lakatos, I. (1970). Falsification and the methodology of scientific research. In I. Lakatos and A. Musgrave (Eds.), *Criticism and the Growth of Knowledge*, ISBN 0521096235, pp. 91-195.

Lamborn, A. C. (1997). Theory and politics in world politics. *International Studies Quarterly*, ISSN 0020-8833, 41(2): 187-214.

Legro, J. W. (1997). Which norms matter? Revisiting the "failure" of internationalism. *International Organization*, ISSN 0020-8183, 51(1): 31-63.

Lempert, R. J. and Schlesinger, M. E. (2000). Robust strategies for abating climate change. *Climatic Change*, ISSN 0165-0009, 45: 387-401.

Lijphart, A. (1975). The comparable-cases strategy in comparative research. *Comparative Political Studies*, ISSN 1552-3829, 8(2): 158-177.

Lindblom, C. E. (2010). The science of "muddling" through. *Emergence: Complexity & Organization*, ISSN 1521-3250, 12(1): 70-80.

Loundes, J. (1997). A brief overview of unemployment in Australia. *Melbourne Institute Working Paper No. 24/97 (page 17)*, ISSN 1328-4991.

Lyotard, J.-F. (1984). *The Postmodern Condition: A Report on Knowledge* (G. Bennington and B. Massumi, Trans.), ISBN 0816611661.

MacIntosh, R. and MacLean, D. (1999). Conditioned emergence: A dissipative structures approach to transformation. *Strategic Management Journal*, ISSN 0143-2095, 20(4): 297-316.

Martel, W. C. (2007). *Victory in War: Foundations of Modern Military Policy*, ISBN 9780521895961.

Mathieson, G. (2004). Full spectrum analysis: Practical OR in the face of the human variable. *Emergence: Complexity and Organization*, ISSN 1521-3250, 6(4): 51-57.

McCarthy, I. P. (2004). Manufacturing strategy: Understanding the fitness landscape. *International Journal of Operations and Production Management*, ISSN 0144-3577, 24(1/2): 124-150.

McIntosh, S. (2007). *Integral Consciousness and the Future of Evolution*, ISBN 9781557788672.

Meehl, P. E. (1992). Cliometric metatheory: The actuarial approach to empirical, history-based philosophy of science. *Psychological Reports*, ISSN 0033-2941, 71: 339-467.

Meehl, P. E. (2002). Cliometric metatheory: II. Criteria scientists use in theory appraisal and why it is rational to do so. *Psychological Reports*, ISSN 0033-2941, 91: 339-404.

Meehl, P. E. (2004). Cliometric metatheory III: Peircean consensus, verisimilitude and asymptotic method. *The British Journal for the Philosophy of Science*, ISSN 0007-0882, 55(4): 615-643.

Menken, H. L. (2009). "Quotations," http://thinkexist.com/ quotation/for_every_problem_there_is_a_solution_which_is/11029.html

Metcalfe, M. (2004). Theory: Seeking a plain English explanation. *Journal of Information Technology Theory and Application*, ISSN 1552-4520, 6(2): 13-21.

Millis, W. (1981). *Arms and Men: A Study in American Military History*, ISBN 9780813509310.

Minati, G. (2010). The dynamic usage of models (DYSAM) as a theoretically-based phenomenological tool for managing complexity and as a research framework. In S. E. Wallis (Ed.), *Cybernetic and Systems Theory in Management: Tools, Views, and Advancements*: 176-190. ISBN 9781615206681.

Nagel, S. S. (Ed.). (1994a). *Encyclopedia of Policy Studies*. (Vol. 13), ISBN 0824791428.

Nagel, S. S. (1994b). Projecting trends in public policy. In S. S. Nagel (Ed.), *Encyclopedia of Policy Studies*: 879-913. ISBN 0824791428.

RateInflation. (2011). "Australia - historical inflation rates," http://www.rateinflation.com/inflation-rate/australia-historical-inflation-rate.php?form=ausir

Rhodes, M. L. (2008). Agent-based modeling for public service policy development: A new framework for policy development. In L. Dennard, K. A. Richardson and G. Morçöl (Eds.), *Complexity and Policy Analysis: Tools and Concepts for Designing Robust Policies in a Complex World*. ISBN 9780981703220, pp. 357-376.

Rhodes, M. L. and Donnelly-Cox, G. (2009). Social entrepreneurship as a performance landscape: The case of 'Front Line'. In J. A. Goldstein, J. K. Hazy and J. Silberstang (Eds.), *Complexity science and social entrepreneurship: Adding social value through systems thinking*. ISBN 978-0-9842164-0-6, pp. 559-580.

Richardson, K. A. (2007). Complex systems thinking and its implications for policy analysis. In G. Morçöl (Ed.), *Handbook of Decision Making*, ISBN 9781574445480, pp. 189-221.

Ritzer, G. (1990). Metatheorizing in sociology. *Sociological Forum*, ISSN 0884-8971, 5(1): 3-15.

Ritzer, G. (2001). *Explorations in Social Theory: From Metatheorizing to Rationalization*, ISBN 0761967737.

Rodrik, D. (1996). Understanding economic policy reform. *Journal of Economic Literature*, ISSN 0022-0515, 34(1): 9-41.

Roe, E. (1998). *Taking Complexity Seriously: Policy Analysis, Triangulation and Sustainable Development*, ISBN 0792380584.

Ross, S. N. and Glock-Grueneich, N. (2008). Growing the field: The institutional, theoretical, and conceptual maturation of "public participation," part 3: Theoretical maturation. *International Journal of Public Participation*, ISSN 1992-6707, 2(1): 14-25.

Sabatier, P. A. (Ed.). (1999). *Theories of the Policy Process*, ISBN 0813399866.

Schlager, E. (1999). Comparison of frameworks, theories, and models of policy processes. In P. A. Sabatier (Ed.), *Theories of the Policy Process:*, Vol. 1: 233-260. ISBN 0-8133-9986-6.

Schmidt, R. E., Scanlon, J. W. and Bell, J. B. (1979). *Evaluability Assessment: Making Public Programs Work Better*, Department of Health Education & Welfare, Project Share.

Schrijver, N. J. (2006). The future of the Charter of the United Nations. *Max Planck Yearbook of United Nations Law*, ISSN 1389-4633, 10: 1-34.

Schuettinger, R. L. and Butler, E. F. (1979). *Forty Centuries of Wage and Price Controls: How Not to Fight Inflation*, ISBN 0891950230.

Scott Jr., R. J. (2010). The science of muddling through revisited. *Emergence: Complexity & Organization*, ISSN 1521-3250, 12(1): 5-18.

Semler, S. (2001). An overview of content analysis. *Practical Assessment, Research & Evaluation*, ISSN 1531-7714, 7(17).

Shane, M. (2010). Real historical gross domestic product (GDP) and growth rates of GDP, www.ers.usda.gov/data/macroeconomics/data/historicalrealgdpvalues.xls

Shotter, J. (2005). Inside the moment of managing: Wittgenstein and the everyday dynamics of our expressive-responsive activities. *Organization Studies*, ISSN 0170-8406, 26(1): 113-135.

Simons, B. A. and Elkins, Z. (2004). The globalization of liberalization: Policy diffusion in the international political economy. *American Political Science Review*, ISSN 003-0554, 98(1): 171-189.

Singer, D. J. (1959). The finances of the League of Nations. *International Organization*, ISSN 0020-8183, 13(2): 255-273.

Smith, M. E. (2003). Changing an organisation's culture: Correlates of success and failure. *Leadership & Organization Development Journal*, ISSN 0143-7739, 24(5): 249-261.

Smolin, L. (2006). *The Trouble with Physics: The Rise of String Theory, the Fall of a Science, and What Comes Next*, ISBN 0618551050.

Sobel, R. S. (1994). The League of Nations Covenant and the United Nations Charter: An analysis of two international constitutions. *Constitutional Political Economy*, ISSN 1043-4062, 5(2): 173-192.

Spicer, M. W. (1998). Public administration, social science, and political association. *Administration and Society*, ISSN 0095-3997, 30(1): 35-53.

Starbuck, W. H. (2003). Shouldn't organization theory emerge from adolescence? *Organization*, ISSN 1350-5084, 10(3): 439-452.

Stinchcombe, A. L. (1987). *Constructing social theories*, ISBN 0-226-77484-8.

Thornton, P. H. and Ocasio, W. (2008). Institutional logics. In R. Greenwood, C. Oliver, S. K. Andersen and R. Suddaby (Eds.), *Handbook of organizational institutionalism*: 99-129. ISBN 9781412931236.

Thornton, P. H. and Ocasio, W. (2008). Institutional logics. In R. Greenwood, C. Oliver, S. K. Andersen and R. Suddaby (Eds.), *Handbook of Organizational Institutionalism*, ISBN 9781412931236, pp. 99-129.

Tsŭ, S. (1981 – original circa 490 BCE). *The Art of War* (L. Giles, Trans.), ISBN 0340276045.

Ugiabe, E. O. and Obetoh, G. (2011). Meta theoretical orientation in social work practice: The relevance of psychodynamics and social ecology theories to social work practice in Nigeria. *Interdisciplinary Journal of Contemporary Research in Business*, ISSN 2073-7122, 3(2): 1346-1355.

United Nations. (1982). "Demographic Yearbook: 1982," http://unstats.un.org/unsd/demographic/products/dyb/dybsets/1982%20DYB.pdf

Van de Ven, A. H. (2007). *Engaged Scholarship: A Guide for Organizational and Social Research*, ISBN 9780199226306.

van Ours, J. C. (2002). Has the Dutch Miracle Come to an End?, paper presented at the *CESifo Conference on Unemployment in Europe*, Munich, Germany.

von Clausewitz, C. (2008 – original 1832). *On War* (J. J. Graham, Trans.), ISBN 1420931822.

Wallis, S. E. (2008a). From reductive to robust: Seeking the core of complex adaptive systems theory. In A. Yang and Y. Shan (Eds.), *Intelligent Complex Adaptive Systems,* ISBN 9781599047171, pp. 1-25.

Wallis, S. E. (2008b). "The integral puzzle: Determining the integrality of integral theory," Essay. http://www.integralworld.net/wallis.html

Wallis, S. E. (2009a). Seeking the robust core of organisational learning theory. *International Journal of Collaborative Enterprise,* ISSN 1740-2085, 1(2): 180-193.

Wallis, S. E. (2009b). Seeking the robust core of social entrepreneurship theory. In J. A. Goldstein, J. K. Hazy and J. Silberstang (Eds.), *Social Entrepreneurship & Complexity.* ISBN 9780984216406, pp. 83-106.

Wallis, S. E. (2010a). The structure of theory and the structure of scientific revolutions: What constitutes an advance in theory? In S. E. Wallis (Ed.), *Cybernetics and Systems Theory in Management: Views, Tools, and Advancements*: 151-174. ISBN 9781615206681.

Wallis, S. E. (2010b). Toward a science of metatheory. *Integral Review,* ISSN 1553-3069, 6(Special Issue: Emerging Perspectives of Metatheory and Theory).

Wallis, S. E. (2010c). Towards developing effective ethics for effective behavior. *Social Responsibility Journal,* ISSN 1747-1117, 6(4): 536-550.

Wallis, S. E. (2010d). Towards the development of more robust policy models. *Integral Review*, ISSN 1554-0790, 6(1): 153-160.

Wallis, S. E. (2011a). The complexity of complexity theory: An innovative analysis. In P. M. Allen, K. A. Richardson and J. A. Goldstein (Eds.), *Emergence, Complexity and Organization: E:CO Annual*, Vol. 11: 179-200. ISBN 978-0-9842165-6-7.

Wallis, S. E. (2011b). From reductive to robust: Seeking the core of institutional theory. *Under submission - available upon request*.

Walters, F. P. (1986). *A History of the League of Nations*, ISBN 9780313250569.

Walters, L. C., Aydelotte, J. and Miller, J. (2000). Putting more public in policy analysis. *Public Administration Review*, ISSN 0033-3352, 60(4): 349-359.

Wawro, G. (2003). *The Franco-Prussian War: The German Conquest of France in 1870-1871*, ISBN 0521584361.

Weick, K. E. (1989). Theory construction as disciplined imagination. *Academy of Management Review*, ISSN 0363-7425, 14(4): 516-531.

Weigley, R. F. (1973). *The American Way of War: A History of United States Military Strategy and Policy*, ISBN 025328029x.

Welters, R. and Muysken, J. (2004). The Philips employment scheme, working paper for the *Centre of Full Employment and Equity Europe*, Maastricht University.

White, L. G. (1994). Values, ethics, and standards in policy analysis. In S. S. Nagel (Ed.), *Encyclopedia of Policy Studies*, 2 ed., ISBN 0-8247-9142-8, pp. 857-878.

Wight, C. (2011). Philosophy of social science and international relations. In W. Carlsnaes, T. Risse and B. Simmons (Eds.), *Handbook of International Relations*, 2 ed., ISBN 9780761963059, pp. 23-51.

Wollmershäuser, T. (2003). Should central banks react to exchange rate movements? An analysis of the robustness of simple policy rules under exchange rate uncertainty, paper presented at the *Second Workshop on Macroeconomic Policy Research*, Budapest, Hungary.

Wroughton, L. and Kaiser, E. (2008). "Financial storm tips world toward recession," http://www.reuters.com/artic lePrint?articleId=USTRE49763N20081008

Yan, A. and Gray, B. (1994). Bargaining power, management control, and performance in United States--China joint ventures: A comparative case study. *Academy of Management Journal*, ISSN 0001-4273, 37(6): 1478-1517.

Lightning Source UK Ltd.
Milton Keynes UK
UKOW030228301011

181181UK00002B/2/P